WITHDRAWN

Benchmark Papers in Geology

Series Editor: Rhodes W. Fairbridge
Columbia University

Volume

1 ENVIRONMENTAL GEOMORPHOLOGY AND LANDSCAPE CONSERVATION, Volume I: Prior to 1900 / Donald R. Coates
2 RIVER MORPHOLOGY / Stanley A. Schumm
3 SPITS AND BARS / Maurice L. Schwartz
4 TEKTITES / Virgil E. Barnes and Mildred A. Barnes
5 GEOCHRONOLOGY: Radiometric Dating of Rocks and Minerals / C. T. Harper
6 SLOPE MORPHOLOGY / Stanley A. Schumm and M. Paul Mosley
7 MARINE EVAPORITES: Origin, Diagenesis, and Geochemistry / Douglas W. Kirkland and Robert Evans
8 ENVIRONMENTAL GEOMORPHOLOGY AND LANDSCAPE CONSERVATION, Volume III: Non-Urban / Donald R. Coates
9 BARRIER ISLANDS / Maurice L. Schwartz
10 GLACIAL ISOSTASY / John T. Andrews
11 GEOCHEMISTRY OF GERMANIUM / Jon N. Weber
12 ENVIRONMENTAL GEOMORPHOLOGY AND LANDSCAPE CONSERVATION, Volume II: Urban Areas / Donald R. Coates
13 PHILOSOPHY OF GEOHISTORY: 1785–1970 / Claude C. Albritton, Jr.
14 GEOCHEMISTRY AND THE ORIGIN OF LIFE / Keith A. Kvenvolden
15 SEDIMENTARY ROCKS: Concepts and History / Albert V. Carozzi
16 GEOCHEMISTRY OF WATER / Yasushi Kitano
17 METAMORPHISM AND PLATE TECTONIC REGIMES / W. G. Ernst
18 GEOCHEMISTRY OF IRON / Henry Lepp
19 SUBDUCTION ZONE METAMORPHISM / W. G. Ernst
20 PLAYAS AND DRIED LAKES: Occurrence and Development / James T. Neal
21 GLACIAL DEPOSITS / Richard P. Goldthwait
22 PLANATION SURFACES: Peneplains, Pediplains, and Etchplains / George F. Adams
23 GEOCHEMISTRY OF BORON / C. T. Walker
24 SUBMARINE CANYONS AND DEEP-SEA FANS: Modern and Ancient / J. H. McD. Whitaker
25 ENVIRONMENTAL GEOLOGY / Frederick Betz, Jr.
26 LOESS: Lithology and Genesis / Ian J. Smalley

27 PERIGLACIAL PROCESSES / *Cuchlaine A. M. King*
28 LANDFORMS AND GEOMORPHOLOGY: Concepts and History / *Cuchlaine A. M. King*
29 METALLOGENY AND GLOBAL TECTONICS / *Wilfred Walker*
30 HOLOCENE TIDAL SEDIMENTATION / *George deVries Klein*
31 PALEOBIOGEOGRAPHY / *Charles A. Ross*
32 MECHANICS OF THRUST FAULTS AND DÉCOLLEMENT / *Barry Voight*
33 WEST INDIES ISLAND ARCS / *Peter H. Mattson*
34 CRYSTAL FORM AND STRUCTURE / *Cecil J. Schneer*
35 OCEANOGRAPHY: Concepts and History / *Margaret B. Deacon*
36 METEORITE CRATERS / *G. J. H. McCall*
37 STATISTICAL ANALYSIS IN GEOLOGY / *John M. Cubitt and Stephen Henley*
38 AIR PHOTOGRAPHY AND COASTAL PROBLEMS / *Mohamed T. El-Ashry*
39 BEACH PROCESSES AND COASTAL HYDRODYNAMICS / *John S. Fisher and Robert Dolan*
40 DIAGENESIS OF DEEP-SEA BIOGENIC SEDIMENTS / *Gerrit J. van der Lingen*
41 DRAINAGE BASIN MORPHOLOGY / *Stanley A. Schumm*
42 COASTAL SEDIMENTATION / *Donald J. P. Swift and Harold D. Palmer*
43 ANCIENT CONTINENTAL DEPOSITS / *Franklyn B. Van Houten*
44 MINERAL DEPOSITS, CONTINENTAL DRIFT AND PLATE TECTONICS / *J. B. Wright*
45 SEA WATER: Cycles of the Major Elements / *James I. Drever*
46 PALYNOLOGY, PART I: Spores and Pollen / *Marjorie D. Muir and William A. S. Sarjeant*
47 PALYNOLOGY, PART II: Dinoflagellates, Acritarchs, and Other Microfossils / *Marjorie D. Muir and William A. S. Sarjeant*
48 GEOLOGY OF THE PLANET MARS / *Vivien Gornitz*
49 GEOCHEMISTRY OF BISMUTH / *Ernest E. Angino and David T. Long*
50 ASTROBLEMES—CRYPTOEXPLOSION STRUCTURES / *G. J. H. McCall*
51 NORTH AMERICAN GEOLOGY: Early Writings / *Robert M. Hazen*

Benchmark Papers in Geology/51

A BENCHMARK® Books Series

NORTH AMERICAN GEOLOGY
Early Writings

Edited by
ROBERT M. HAZEN
Carnegie Institution of Washington

STROUDSBURG, PENNSYLVANIA

Copyright © 1979 by **Dowden, Hutchinson & Ross, Inc.**
Benchmark Papers in Geology, Volume 51
Library of Congress Catalog Card Number: 79-708
ISBN: 0-87933-345-6

All rights reserved. No part of this book covered by the copyrights hereon may be reproduced or transmitted in any form or by any means—graphic, electronic, or mechanical, including photocopying, recording, taping, or information storage and retrieval systems—without written permission of the publisher.

81 80 79 1 2 3 4 5
Manufactured in the United States of America.

LIBRARY OF CONGRESS CATALOGING IN PUBLICATION DATA
Main entry under title:
North American Geology: Early Writings
  (Benchmark papers in geology ; 51)
  Bibliography: p.
  Includes indexes.
  1. Geology—North America—Addresses, essays, lectures. I. Hazen, Robert M., 1948-
QE71.E19   557'.08   79-708
ISBN: 0-87933-345-6

Distributed world wide by Academic Press,
a subsidiary of Harcourt Brace Jovanovich,
Publishers.

SERIES EDITOR'S FOREWORD

The philosophy behind the "Benchmark Papers in Geology" is one of collection, sifting, and rediffusion. Scientific literature today is so vast, so dispersed, and, in the case of old papers, so inaccessible for readers not in the immediate neighborhood of major libraries that much valuable information has been ignored by default. It has become just so difficult, or so time consuming, to search out the key papers in any basic area of research that one can hardly blame a busy person for skimping on some of his or her "homework."

This series of volumes has been devised, therefore, to make a practical contribution to this critical problem. The geologist, perhaps even more than any other scientist, often suffers from twin difficulties—isolation from central library resources and immensely diffused sources of material. New colleges and industrial libraries simply cannot afford to purchase complete runs of all the world's earth science literature. Specialists simply cannot locate reprints or copies of all their principal reference materials. So it is that we are now making a concerted effort to gather into single volumes the critical materials needed to reconstruct the background of any and every major topic of our discipline.

We are interpreting "geology" in its broadest sense: the fundamental science of the planet Earth, its materials, its history, and its dynamics. Because of training in "earthy" materials, we also take in astrogeology, the corresponding aspect of the planetary sciences. Besides the classical core disciplines such as mineralogy, petrology, structure, geomorphology, paleontology, and stratigraphy, we embrace the newer fields of geophysics and geochemistry, applied also to oceanography, geochronology, and paleoecology. We recognize the work of the mining geologists, the petroleum geologists, the hydrologists, and the engineering and environmental geologists. Each specialist needs a working library. We are endeavoring to make the task of compiling such a library a little easier.

Each volume in the series contains an introduction prepared by a specialist (the volume editor)—a "state of the art" opening or a summary of the object and content of the volume. The articles, usually some twenty to fifty reproduced either in their entirety or in significant extracts, are selected in an attempt to cover the field, from the key papers of the last century to fairly recent work. Where the original works are in foreign

Series Editor's Foreword

languages, we have endeavored to locate or commission translations. Geologists, because of their global subject, are often acutely aware of the oneness of our world. The selections cannot, therefore, be restricted to any one country, and whenever possible an attempt is made to scan the world literature.

To each article, or group of kindred articles, some sort of "highlight commentary" is usually supplied by the volume editor. This commentary should serve to bring that article into historical perspective and to emphasize its particular role in the growth of the field. References, or citations, wherever possible, will be reproduced in their entirety—for by this means the observant reader can assess the background material available to that particular author, or, if desired, he or she, too, can double check the earlier sources.

A "benchmark," in surveyor's terminology, is an established point on the ground, recorded on our maps. It is usually anything that is a vantage point, from a modest hill to a mountain peak. From the historical viewpoint, these benchmarks are the bricks of our scientific edifice.

RHODES W. FAIRBRIDGE

PREFACE

The emergence of North American geology is a fascinating branch of the history of science. Geology was closely linked with the discovery, exploration, and expansion of the continent, and the growth of this discipline reflects the changing religious, social, and educational climate of the New World. In spite of a growing interest and awareness in all aspects of the history of science, American earth science before 1820 has received less analysis than subsequent decades. George P. Merrill (1924), for example, considered *The First One Hundred Years of American Geology* to be 1785 to 1885, but he devoted only one of fifteen chapters to the years before 1820. This relative neglect is not surprising, for the early literature on North American earth phenomena is little known and inaccessible to many historians of science. Physicians, clergymen, and a few well-educated, wealthy gentlemen were the first writers on geology and other sciences in this country; consequently the published literature on earth science is widely scattered in travel accounts, medical periodicals, privately published pamphlets, and proceedings of literary societies. Many of these early American sources are all but forgotten, and few pre-1820 contributions have been reprinted.

Hazen and Hazen (1976) recently compiled the *Bibliography of American-Published Geology*, which contains 14,000 references on geology for the period 1699 to 1850. The nearly 2,000 articles, books, maps, and pamphlets from before 1821 form the basis for reprint selections in the present volume. Early American publications in this volume were chosen on the basis of author, title, and source. An effort was made to include at least one contribution by each major American earth scientist from the eighteenth and first decades of the nineteenth centuries. Articles were also chosen to illustrate eight topical sections representative of the range of geological subjects considered by naturalists prior to 1820. It should be noted, however, that these eight sections are used as a convenience in presenting a large and diverse literature, and the topics do not necessarily reflect an historical division of the earth sciences. A third criterion for article selection was the original source. Early American publications in geology appeared in dozens of obscure and short-lived periodicals; these geological contributions have often been forgotten along with the source. Several articles from these little-known and rare journals are reprinted in

Preface

this volume. Many articles of historical and scientific interest could not be included because of space limitations, so primary-source bibliographies have been included at the end of each section to direct the reader to additional important publications. A bibliography of secondary sources on the history of geology, particularly in North America, is included at the end of the Introduction.

This Benchmark volume contains a wide range of articles on topics that are now recognized under the heading of geology. Not included here are the contributions of North American travelers and explorers, whose accounts include only scattered references to soils and rocks. Descriptions of early American mines and mining will appear in another Benchmark volume. The important early field observations by the Hessian doctor Johann David Schöpf (Spieker 1972) and the Frenchman C. F. C. Comte de Volney (White 1968) have been reprinted and discussed in detail elsewhere.

All reprints in this volume were originally printed in North America, and all but two were written by Americans. This emphasis on North American geology does not imply a lack of influence by, or interaction with, European earth scientists. Examination of these American contributions, on the contrary, reveals both admiration for and dependence upon European efforts. The science of geology, as we know it today, was largely a European development. Americans before 1820 followed the lead of their trans-Atlantic neighbors, and much of North American progress during this period may be viewed as an effort to emulate European understanding of earth history. The leadership of European earth scientists, therefore, is reflected in American writings.

All reprints were photographed directly from original sources, some of which are worn or stained. It is the editor's belief that this reproduction of original format and type is more desirable than the slight increase in legibility which might result from resetting the material.

It is a pleasure to acknowledge the aid and advice of Brooke Hindle, whose writings, teachings, and discussions on the history of American science were instrumental in the preparation of this volume. The constructive review by Cecil Schneer improved the work, as did the perceptive comments of George White on the early literature of American earth science. Margaret Hazen assisted in historical analysis of the sources, and Dan C. Hazen and Dolores M. Thomas suggested numerous valuable editorial revisions. Thanks are also due to the many librarians who made rare early texts available for study. John Blake of the National Medical Library, Bethesda, Maryland, and the staff of the United States Geological Survey Library, Reston, Virginia, were especially helpful. Grateful appreciation is also extended to the Georgia Historical Society for providing photographs of the original material included as Paper 3, to the Medical Research Library of the Brooklyn Downstate Medical Center for the loan of the original volume from which Paper 9 was reproduced, to the Library of the New York Botanical Garden, Bronx, New York, for lending the volume from which Paper 19 was reproduced, and to J. S. Canner & Company for lending the volume from which Paper 29 was reproduced.

ROBERT M. HAZEN

REFERENCES

Hazen, R. M., and M. H. Hazen. 1976. *Bibliography of American-Published Geology: 1669 to 1850.* Geol. Soc. Amer., Microform Publ. #4.

Merrill, G. P. 1924. *The First One Hundred Years of American Geology.* New Haven, Conn.: Yale Univ. Press.

Spieker, E. M. 1972. *Geology of Eastern North America by Johann David Schöpf. An Annotated Translation.* . . .New York: Hafner.

White, G. W. 1968. "Introduction" to a reprint edition of C. F. Volney's *View of the Soil and Climate of the United States.* New York: Hafner.

CONTENTS

Series Editor's Foreword — vii
Preface — ix
Contents by Author — xvii

Introduction — 1

PART I: EARTHQUAKES

Editor's Comments on Papers 1 and 2 — 10

1 PRINCE, THOMAS: Earthquakes, the Works of God and Tokens of His Just Displeasure — 13
Boston: D. Henchman, 1727, pp. i, iii, v–vi, 1–20

2 WINTHROP, JOHN: A Lecture on Earthquakes — 37
Boston: Edes and Gill, 1755, pp. i, iii, 5–31

PART II: FOSSILS

Editor's Comments on Papers 3 Through 7 — 68

3 ANONYMOUS: Description of a Remarkable Tooth, in the Possession of Mr. Peale — 73
The Columbian Mag., or Mon. Misc. 1:655 (1787)

4 JEFFERSON, THOMAS: A Memoir on the Discovery of Certain Bones of a Quadruped of the Clawed Kind in the Western Parts of Virginia — 74
Am. Philos. Soc. Trans. 4:246–258 (1799)

5 BARTON, BENJAMIN SMITH: Facts, Observations, and Conjectures, Relative to the Elephantine Bones (of Different Species), That Are Found in Various Parts of North America — 87
Phila. Med. Phys. Jour. **1st suppl.**:22–35 (1806)

6 CLEAVELAND, PARKER: Account of Fossil Shells, with the Author's Reasons for Attending to the Same — 101
Am. Acad. Arts Sci. Mem. 3, pt. 1, 155–158 (1809)

7 LE SUEUR, CHARLES ALEXANDER: Observations on a New Genus of Fossil Shells — 105
Phila. Acad. Nat. Sci. Jour. 1, pt. 2, 310–312 (1818)

Contents

PART III: MEDICAL GEOLOGY

Editor's Comments on Papers 8 and 9   110

   8   MITCHILL, SAMUEL LATHAM: Outlines of Medical Geography: Being an Inquiry How Far Calcareous Soils and Strata Counteract the Septic Exhalations Which Occasion Distempers of a Febrile or Pestilential Type   113
Med. Repos. **2**:39–47 (1799)

   9   ANONYMOUS: Reviews of Six Treatises on Mineral Springs of the United States   122
New England Jour. Med. Surg. **6**:363–385 (1817)

PART IV: MINERALOGY

Editor's Comments on Papers 10, 11, and 12   146

   10   ANONYMOUS: Mineralogy   152
Encyclopedia, or a Dictionary of Arts, Sciences, and Miscellaneous Literature, Vol. 12, Phila., Pa.: T. Dobson, 1794, pp. 58–69

   11   BRUCE, ARCHIBALD: Description, and Chemical Examination of an Ore of Zinc, from New Jersey   165
Am. Mineralog. Jour. **1**:96–99 (1811)

   12   SILLIMAN, BENJAMIN: Review of An Elementary Treatise on Mineralogy and Geology Being an Introduction to the Study of These Sciences, and Designed for the Use of Pupils; for Persons Attending Lectures on These Subjects; and As a Companion for Travellers in the United States of America—Illustrated by Six Plates, by Parker Cleaveland   169
Am. Jour. Sci. **1**:35–52 (1818)

PART V: PHYSICS OF THE EARTH

Editor's Comments on Papers 13 Through 16   188

   13   FRANKLIN, BENJAMIN: Conjectures Concerning the Formation of the Earth   192
Am. Philos. Soc. Trans. **3**:1–5 (1793)

   14   FRANKLIN, BENJAMIN: Queries and Conjectures Relating to Magnetism, and the Theory of the Earth   197
Am. Philos. Soc. Trans. **3**:10–13 (1793)

   15   COOPER, THOMAS: Geology   201
Emporium Arts Sci. n.s. **3**:412–425 (1813)

   16   BOWDITCH, NATHANIEL: On the Calculation of the Oblateness of the Earth, by Means of the Observed Lengths of a Pendulum in Different Latitudes, According to the Method Given by La Place in the Second Volume of His "Mécanique Céleste," with Remarks on Other Parts of the Same Work, Relating to the Figure of the Earth   215
Am. Acad. Arts Sci. Mem. **4**, pt. 1, 30–34 (1818)

PART VI: FORMATION AND CLASSIFICATION OF ROCKS

Editor's Comments on Papers 17 Through 22 — 222

17 SMITH, THOMAS P.: Account of Chrystallized Basaltes Found in Pennsylvania — 228
  Am. Philos. Soc. Trans. **4**:445–446 (1799)

18A HALL, REVEREND JAMES: An Account of a Supposed Artificial Wall, Discovered Under the Surface of the Earth in North Carolina — 230
  Med. Repos. **2**:272–274 (1799)

18B WOODHOUSE, JAMES: Reply — 233
  Med. Repos. **2**:275–278 (1799)

19 LEWIS, ZECHARIAH: Remarks on a Subterranean Wall in North Carolina — 237
  Med. Repos. **4**:227–232 (1801)

20 MACLURE, WILLIAM: Essay on the Formation of Rocks, or an Inquiry into the Probable Origin of Their Present Form and Structure — 243
  Phila. Acad. Nat. Sci. Jour. **1**:261–271 (1818)

21 MACLURE, WILLIAM: Hints on Some of the Outlines of Geological Arrangements, with Particular Reference to the System of Werner, and with Introductory Remarks by Benjamin Silliman — 254
  Am. Jour. Sci. **1**:209–213 (1819)

22 DAY, JEREMIAH: A View of the Theories Which Have Been Proposed, to Explain the Origin of Meteoric Stones — 259
  Connecticut Acad. Arts Sci. Mem. **1**, pt. 1, 163–174 (1810)

PART VII: FIELD GEOLOGY

Editor's Comments on Papers 23 Through 28 — 272

23 MITCHILL, SAMUEL LATHAM: Geological Remarks on Certain Maritime Parts of the State of New York — 279
  Am. Mus. or Univers. Mag. **5**:123–126 (1789)

24 MACLURE, WILLIAM: Observations on the Geology of the United States, Explanatory of a Geological Map — 283
  Am. Philos. Soc. Trans. **6**, pt. 2, 411–428 (1809)

25 MITCHILL, SAMUEL LATHAM: An Amendment Proposed to the Geological Chart of the United States, as Respects the Character of the North Side of Long Island, Which Is Shown to be Alluvial and Not Primitive, as Therein Stated — 301
  Am. Mineralog. Jour. **1**, pt. 2, 129–133 (1811)

26 RAFINESQUE, C. S.: Review of Observations on the Geology of the United States of America; With Some Remarks on the Effects Produced on the Nature and Fertility of Soils, by

Contents

	the Decomposition of the Different Classes of Rocks, and an Application to the Fertility of Every State in the Union, in Reference to the Accompanying Geological Map—With Two Plates, by William Maclure *Am. Monthly Mag. and Crit. Rev.* 3:41–44 (1818)	306
27	RAFINESQUE, C. S.: Review of An Index to the Geology of the Northern States, with a Transverse Section from the Catskill Mountains to the Atlantic, Prepared for the Geological Classes at William's College, Massachusetts, by Amos Eaton *Am. Monthly Mag. and Crit. Rev.* 3:175–178 (1818)	310
28	HAYDEN, H. H.: Agenda, or Selection of Queries *Geological Essays; or, an Inquiry Into Some of the Geological Phenomena to be Found in Various Parts of America, and Elsewhere,* Baltimore, Md.: J. Robinson, 1820, pp. 395–412	314

PART VIII: AMERICAN GEOLOGY COMES OF AGE

Editor's Comments on Papers 29 and 30		334
29	THE EDINBURGH REVIEW: Reviews of *Observations on the Geology of the United States* by William Maclure, *and* An Elementary Treatise on Mineralogy and Geology by Parker Cleaveland *Analectic Mag.* 13:322–326 (1819)	336
30	AMERICAN GEOLOGICAL SOCIETY: Constitution, Officers, and Proceedings for 1819 and 1820 *Am. Jour. Sci.* 2:139–144 (1820)	341

Author Citation Index	347
Subject Index	349
About the Editor	357

CONTENTS BY AUTHOR

American Geological Society, 341
Barton, B. S., 87
Bowditch, N., 215
Bruce, A., 165
Cleaveland, P., 101
Cooper, T., 201
Day, J., 259
Edinburgh Review, 336
Franklin, B., 192, 197
Hall, J., 230
Hayden, H. H., 314

Jefferson, T., 74
Le Sueur, C. A., 105
Lewis, Z., 237
Maclure, W., 243, 254, 283
Mitchill, S. L., 113, 279, 301
Prince, T., 13
Rafinesque, C. S., 306, 310
Silliman, B., 169
Smith, T. P., 228
Winthrop, J., 37
Woodhouse, J., 233

NORTH AMERICAN GEOLOGY

INTRODUCTION

Scientific progress is a complex phenomenon, controlled by numerous and interrelated factors. Dominant personalities and prescient thinkers are of obvious importance in the advancement of any field of human endeavor. Groups of individuals working together may represent a "critical mass" and thus spur rapid advances, while individuals in competition are also vital to scientific progress. No less important in examining the history of science is the social climate in which scientists worked. Religious beliefs, attitudes toward education, conditions of warfare, and economic development have a profound effect on a society's desire and ability to engage in pure and applied research. Even such natural agencies as weather, geographical location, or topography may advance or retard the pursuit of science. All these influences affected the development of the geological sciences in America.

This volume emphasizes geological publications, but progress in the earth sciences cannot be measured by publications alone. Significant contributions were made by educators such as Amos Eaton, Edward Hitchcock, and Benjamin Silliman. Thomas Cooper, Archibald Bruce, Constantine Rafinesque, and Samuel Mitchill advanced geology as editors of scientific journals. Others, including Benjamin Franklin and William Maclure, inspired and encouraged their colleagues through their voluminous correspondence. Collectors spurred geological research by preserving mineral and fossil specimens for study. Interactions with European naturalists provide yet another measure of the development of American geology. The papers reprinted here, therefore, must be viewed in conjunction with the less tangible contributions of teachers, editors, collectors, and others in analyzing the progress of American earth science.

Introduction

The first recorded observations of solid earth phenomena in North America were made by European explorers and by the first settlers of the New World. Gold and other treasures lured Columbus and subsequent voyagers, and early travel accounts contained notes on rocks and soils as part of the general description of these western lands. These few geological observations, however, provided only a fragmentary picture of the underlying structure of the continent.

The earliest American publications relating to the earth also dealt with isolated phenomena; the great majority of American-published earth science literature prior to 1776 was on volcanoes and earthquakes. The first interpretive accounts by North American authors were sermons preached and published after the devastating Port Royal, Jamaica, earthquake of 1692 and less destructive shocks near Boston, Massachusetts, in 1703 and 1727. In the Puritan-founded Boston society, earthquakes were universally ascribed to the wrath of God (Prince, Paper 1.) John Winthrop's attempt at a physical explanation of earthquakes (Paper 2), which followed yet another tremor in Boston in 1755, represented the first original earth science contribution by an American.

The discovery of large fossil bones in various parts of North America attracted the attention of many prominent naturalists. These "organic relicts" were noted as early as 1706, and deposits in New York, Pennsylvania, Virginia, and the western territories were documented prior to the Revolutionary War. David Rittenhouse (Paper 3) and Thomas Jefferson (Paper 4) reported finds of unusual vertebrate remains prior to 1800, and the growing number of bone discoveries enabled Benjamin Smith Barton (Paper 5) to speculate on the habits, varieties, distribution, and extinction of elephants in North America. The 1818 publication of Georges Cuvier's *Essay on the Theory of the Earth* further expanded Americans' understanding of the nature and origin of vertebrate fossil remains.

Fossil shells, though less spectacular than large bones, were recorded by several investigators, including Parker Cleaveland (Paper 6) and Charles Alexander Le Sueur (Paper 7). Invertebrate remains, however, were neither widely studied nor used for stratigraphic correlation in North America until the 1830s.

Many naturalists in the eighteenth and early nineteenth centuries were trained in the medical profession, and they applied their knowledge of natural productions to the prevention and treatment of disease. Samuel Latham Mitchill (Paper 8) believed that calcareous rocks and soils neutralized some causes of disease. He thus espoused the establishment of towns on carbonate sediments, and recommended the use of limestone in buildings and roads. Many physicians studied the medicinal properties of mineral springs and waters in Virginia, New York, and other eastern states. The large body of literature by 1817 prompted the

Introduction

New England Journal of Medicine and Surgery to review the six most significant (and in many particulars, contradictory) accounts of the famed mineral springs at Saratoga and Ballston, New York (Paper 9).

Like most of the natural sciences, mineralogy is a field that relies on an unambiguous system of classification and identification. With few teachers and collectors trained in mineral identification, American mineralogy made little progress before 1800. The 1794 American publication of a mineralogy textbook, a translation of the system by Swedish mineralogist and chemist Axel Cronstedt (Paper 10), had almost no impact on the American scientific community, which possessed little of the requisite practical training in mineral identification. The introduction of chemistry courses at American colleges and the efforts of such mineral collectors as Colonel George Gibbs allowed American mineralogical expertise to advance rapidly. By 1810 Archibald Bruce commenced the *American Mineralogical Journal*, which served as a focus for the emerging science in the United States (Paper 11). The swift development of mineralogy was further demonstrated with Parker Cleaveland's *Elementary Treatise on Mineralogy and Geology* (1816), the first such text written by an American. This major effort was the subject of numerous favorable reviews, including Benjamin Silliman's glowing retrospective analysis (Paper 12).

The formation of the earth, though a problem far more speculative than the empirics of mineral or fossil identification, nonetheless inspired publications from at least two early American thinkers. The most noted of these contributions were two letters of conjectures and queries by Benjamin Franklin (Papers 13 and 14). His speculations ranged from the nature of the earth's core and the origin of the earth's magnetic field to the causes and consequences of polar wandering. Several decades later, Thomas Cooper contributed his own thoughts on the nature of the earth's core by suggesting a connection between the composition of the earth's core and meteorites (Paper 15). A more quantitative problem relating to the earth was the shape or "figure" of the globe. An early solution for determining the earth's deviations from a sphere was proposed by the French mathematician Laplace, and this presentation was modified and offered to an American audience by Nathaniel Bowditch (Paper 16).

Geology was understood as a historical science by early American workers, and the origin of rocks was recognized as an important element of that history. The system of the German geologist Abraham Werner, which was based on the concept of a universal sequence of rock deposition in a vast global ocean, was widely accepted in America prior to 1820. Thomas P. Smith (Paper 17) thus viewed jointed basaltic blocks as crystalline masses of Neptunian origin. Certain rock formations seemed more puzzling to early workers, as in the case of a natural

Introduction

"wall" or dike in North Carolina (Papers 18A, 18B, and 19). Chemist James Woodhouse recognized the dike to be a natural production of volcanic origin, but to other workers the wall was unmistakably of human workmanship.

William Maclure, a pivotal figure in North American geology before 1820, initially accepted Wernerian nomenclature for want of a better system. In 1818, however, Maclure proposed a new classification in his "Essay on the formation of rocks" (Paper 20). Maclure's system included classes for both Neptunian or water-formed rocks, in accordance with Werner, and volcanic or fire-formed rocks, as described in the theory of Scottish geologist James Hutton. Maclure thus abandoned much of Werner's nomenclature with its implication of water genesis. Maclure's dissatisfaction with some aspects of the Wernerian system was further emphasized in a brief critique of 1819 (Paper 21) in which he noted that many rocks are not found in the universal sequence Werner proposed. An editorial preface to this note by Benjamin Silliman documented the growing prestige of Maclure and the increasing recognition of shortcomings in the Wernerian system.

Terrestrial formations were not the only rocks to excite the attention of American geologists. Meteorites were a subject of wonder in the early nineteenth century, and a brilliant fall near Weston, Connecticut, in 1806 was no exception. This occurrence was rendered especially significant because of the documentation provided by Benjamin Silliman and James Kingsley, who convinced many skeptics of the reality of meteorites. The origins of meteorites, however, remained in doubt, as suggested by Jeremiah Day's review of several possible theories under consideration shortly after the Weston fall (Paper 22).

Field geology in North America began with the observations of explorers and settlers, but the first systematic investigations into the continent's underlying structure were not commenced until after 1780. Johann Schöpf, a Hessian physician, is usually credited with the first major study of North American geology, but his efforts were published in Germany and did not reach a contemporary American readership. Samuel Latham Mitchill was perhaps the first American to publish a local geological account, "Geological remarks on certain maritime parts of the state of New-York," in 1787 (Paper 23). An important eighteenth-century geological study by a foreign visitor was *A View of the Soil and Climate of the United States* by C. F. Volney, a French naturalist. This study, which contained the first true geological map of North America, successfully outlined the general trend of rock units in the eastern part of the continent.

William Maclure's *Observations on the Geology of the United States*, first published in 1809, was the most important pre-1820 contribution to North American geology (Paper 24). Maclure's *Observations* and the accompanying geological map covered more than half a million square

Introduction

miles of eastern North America and served as a base for future field studies. His essay represented a rough outline of American geology, for it contained only four different rock formations. Subsequent investigations such as S. L. Mitchill's account of Long Island (Paper 25) were successful in correcting errors and adding details to Maclure's foundation. Maclure revised his original observations partly on the basis of additions and corrections by Mitchill and others, and he published the second edition of the *Observations* as a pamphlet in 1817. Maclure's efforts achieved widespread acclaim in American and foreign reviews (Papers 26 and 29), and his essay long remained the standard work on American geology.

Several detailed local studies closely followed Maclure's second *Observations*. The noted geologists Edward Hitchcock and Amos Eaton both published their first geological articles and maps in 1818. In that year Eaton also wrote *An Index to the Geology of the Northern States*, a text on the geological structure of New England and eastern New York with a representative east-west cross-section. This effort was favorably reviewed by C. S. Rafinesque (Paper 27), who particularly recognized the increase in geological detail over Maclure's pioneer study. By 1820 field geology had progressed to the point that Horace Hayden could propose a set of geological queries (Paper 28) that might guide investigators in the systematic study of any area.

Through the efforts of both foreign and American workers, geology had come of age in North America. The final two reprint selections illustrate two aspects of the emergence of American geology. The recognition of American scientific achievement by European leaders in geology was a relatively infrequent, though much desired event in America. The prestigious *Edinburgh Review's* favorable analysis of Maclure's *Observations* and of Cleaveland's *Elementary Treatise* (Paper 29) was thus a source of pride to the American earth science community. This international recognition was nearly simultaneous with the organization in 1819 of the American Geological Society (Paper 30), the first such professional society for American earth scientists. By 1820 geology in North America had gained an international reputation and was an organized and established science. The geologists of North America would play an important role in the continuing exploration, expansion, and development of the New World.

BIBLIOGRAPHY OF SECONDARY SOURCES

Adams, F. D. 1932. Earliest Use of the Term Geology. *Geol. Soc. America Bull.* **43**: 121-123.

Albritton, C. C. 1975. *Philosophy of Geohistory: 1785-1970.* Benchmark Papers in Geology, vol. 13. Stroudsburg, Penn.: Dowden, Hutchinson and Ross, 386p.

Introduction

Aldridge, A. O. 1950. Benjamin Franklin and Jonathan Edwards on Lightning and Earthquakes. *Isis* **41**:162–164.

Bell, W. J. 1939. A Box of Old Bones: A Note on the Identification of the Mastodon, 1766–1806. *Am. Philos. Soc. Proc.* **93**:169–177.

Bell, W. J. 1955. *Early American Science Needs and Opportunities for Study*, vol. 9. Williamsburg, Virg.: Inst. of Early American History and Culture, 85p.

Burke, John G. 1969. Mineral Classification in the Early Nineteenth Century. In *Toward a History of Geology*, edited by C. J. Schneer. Cambridge, Mass.: MIT Press, pp. 62–77.

Fitzpatrick, T. J. 1911. *Rafinesque: A Sketch of His Life, with Bibliography.* Des Moines, Iowa: Historical Dept. of Iowa, 241p.

Frondel, C. 1971. Early Mineral Specimens from New England. *Mineralog. Rec.* **2**:232–234.

Fulton, J. F., and E. H. Thomson. 1947. *Benjamin Silliman—Pathfinder in American Science.* New York: H. Schumann, xiii, 294p.

Greene, J. C. 1954. Some Aspects of American Astronomy, 1750–1815. *Isis* **45**:339–358.

Greene, J. C. 1959. *The Death of Adam, Evolution and Its Impact on Western Thought.* Ames, Iowa: Iowa State Univ. Press, 388p.

Greene, J. C. 1969. The Development of Mineralogy in Philadelphia, 1780–1820. *Am. Philos. Soc. Proc.* **113**:283–295.

Greene, J. C., and J. G. Burke. 1978. The Science of Minerals in the Age of Jefferson. *Am. Philos. Soc. Trans.* **68**, pt. 4, 113p.

Hall, Courtney R. 1934. *A Scientist in the Early Republic: Samuel Latham Mitchill, 1764–1831.* New York: Columbia Univ. Press, 162p.

Hazen, R. M., and M. H. Hazen. 1976. *Bibliography of American-published Geology: 1669 to 1850.* Geol. Soc. Amer., Microform Publ. #4, 979p.

Hazen, R. M., and M. H. Hazen. 1978. Neglected Geological Literature: An Introduction to a Bibliography of American-published Geology. In *Proc. New Hampshire Bicentennial Conf. on the History of Geology*, edited by C. J. Schneer. In Press.

Hindle, B. 1956. *The Pursuit of Science in Revolutionary America.* Chapel Hill: Univ. of North Carolina Press, xiii, 410p.

McAllister, E. M. 1941. *Amos Eaton, Scientist and Educator, 1776–1842.* Philadelphia: Univ. of Pennsylvania Press, 587p.

Meisel, M. 1924–1927. *A Bibliography of American Natural History. The Pioneer Century, 1769–1865.* New York: Premier Pub. Co., 3 volumes.

Merrill, G. P. 1924. *The First One Hundred Years of American Geology.* New Haven, Conn.: Yale Univ. Press, xxi, 773p.

Moore, J. 1947. William Maclure: Scientist and Humanitarian. *Am. Philos. Soc. Proc.* **91**:234–249.

Penick, J., Jr. 1976. *The New Madrid Earthquakes of 1811–1812.* Columbia, Missouri: Univ. of Missouri Press, ix, 181p.

Pomerantz, M. A. 1976. Benjamin Franklin—The Compleat (sic) Geophysicist. *EOS-Am. Geophys. Union Trans.* **57**:492–505.

Rudwick, M. J. S. 1972. *The Meaning of Fossils*, 2d ed. New York: Science History Publications, 287p.

Schneer, C. J., ed. 1969. *Toward a History of Geology.* Cambridge, Mass.: MIT Press, vi, 469p.

Simpson, G. G. 1942. The Beginnings of Vertebrate Paleontology in North America. *Am. Philos. Soc. Proc.* **86**:139.

Simpson, G. G. 1943. The Discovery of Fossil Vertebrates in North America. *Jour. Paleontology* **17**:26–38.

Spieker, E. M. 1972 *Geology of Eastern North America by Johann David Schöpf. An Annotated Translation* . . . New York: Hafner, 171 + 195p.

Spieker, E. M., and R. M. Hazen. 1975. The Founding of Geology in America, 1771 to 1818: Discussion and Reply. *Geol. Soc. America Bull.* **86**:1615-1616.

Wells, J. W. 1963. Notes on the Earliest Geological Maps of the United States, 1756-1832. *Jour. Washington Acad. Sci.* **49**:198-205.

Wells, J. W. 1963. *Early Investigations of the Devonian System in New York, 1656-1836.* New York: Geol. Soc. America Spec. Paper 74, 74p.

White, G. W. 1951. Lewis Evans' Contributions to Early American Geology, 1743-1755. *Illinois Acad. Sci. Trans.* **44**:152-158.

White, G. W. 1967. Account of Illinois Geology by John Bradbury in 1817. *Illinois State Acad. Sci. Trans.* **60**:337-339.

White, G. W. 1969. Early Geological Observations in the American Midwest. In *Toward a History of Geology,* edited by C. J. Schneer. Cambridge, Mass.: MIT Press, pp. 415-425.

White, G. W. 1970. William Maclure Was a Uniformitarian and Not a Real Wernerian. *Jour. Geol. Education* **18**:127-128.

White, G. W. 1968. "Introduction" to a reprint edition of C. F. Volney's *View of the Soil and Climate of the United States.* New York: Hafner.

White, G. W. 1973. The History of Geology and Mineralogy as Seen by American Writers, 1803-1835: A Bibliographic Essay. *Isis* **64**:197-214.

Part I
EARTHQUAKES

Editor's Comments on Papers 1 and 2

1   **PRINCE**
Excerpt from *Earthquakes, the Works of God and Tokens of His Just Displeasure*

2   **WINTHROP**
A Lecture on Earthquakes

Earthquakes, perhaps more than any other physical phenomena, cause man to reflect on his place in nature. To feel the solid earth sway and contort undermines man's confidence in his most stable frame of reference. In 1727 and again in 1755, moderate earthquakes rocked Boston, Massachusetts, and nearby regions. At a time when God's will was implicit in all natural occurrences, these earthquakes caused fear and inspired repentance in the hearts of many New Englanders. It was to the clergy that citizens turned for guidance and support, and the colonial pastors responded with numerous earthquake sermons. More than forty pamphlet sermons, all sounding a common theme, were published in New England shortly after the 1727 and 1755 shocks.

Earthquakes, the preachers stated, were caused by God in his anger against a sinful people. Typically, as in the sermon by Thomas Prince (Paper 1), scriptural accounts and interpretations of earthquakes were primary sources for this point of view. The proclamation of Psalm 18, "then the earth shook . . . because he was wroth," was incontrovertible, and the congregations of New England responded with days of fasting and prayer.

Thomas Prince (1687–1758), like many of his fellow theologians, clearly distinguished between the "first cause" of earthquakes, that is, God's judgment of His people, and the natural or "second cause." That the earthquakes had a physical origin, consistent with natural laws, was not generally denied, although most sermon authors avoided discussing this issue. Prince's exposition is notable for his direct treatment of "natural, instrumental, or secondary causes," which are "the means He uses to accomplish his pleasure" (pp. 9-11). Prince presented the conventional European belief that the earth's interior has "a vast and inconceivable number of caverns or hollow places" that contain vapors.

Explosions or sudden movement of these vapors may disturb the confining cavern walls, and consequently affect the surface of the earth.

> Thus has the most high God ... made sufficient provision for such dire convulsions. He has placed us over great and hideous vaults, large enough to receive the most spacious cities, and ready to open when he sees it time to bury us in them.

These comments on "second causes" were not calculated to reassure the sinner!

An alternative interpretation of the earthquake phenomenon was proposed by John Winthrop (1714-1779), fourth in his distinguished family to bear the name. Winthrop, who counted governors of Massachusetts and Connecticut and the first American member of the Royal Academy among his famous ancestors, was best known for his contributions to astronomy. At a time when precise measurements of astronomical events in widely separated locations were essential to calculations of the earth-to-sun distance, Winthrop's American-based observations were of international importance (Hindle 1956).

Because of his position as Harvard College's Hollis Professor of Mathematics and Philosophy, Winthrop was called upon to give his views on earthquakes after the 1755 Boston tremor. Winthrop delivered his *Lecture on Earthquakes* (Paper 2) in the College Chapel, just eight days after the shock. Unlike previous New Englanders who wrote about earthquakes, Winthrop emphasized second causes and produced the first original earth science contribution by an American. Winthrop commenced his lecture with a detailed description of the phenomena, based on his own and others' observations (pp. 1-16). He proceeded to examine the probable cause of earthquakes and proposed essentially the same mechanism that was discussed by Prince in 1727—namely, that hot gases expanding in subterranean passages disturbed the earth's surface. Winthrop elaborated on this theme by noting for the first time that the movement of underground gases through long passages caused the earthquake to travel in a "progressive swell or undulation."

A major difference between the Winthrop and Prince discussions was the assessment of God's motivation in causing earthquakes. Unlike Prince, Winthrop saw a beneficent God in the New England earthquakes, for no one was seriously injured, and a potential source of devastation was relieved by the escape of the expanding gases. He acknowledged that the most destructive earthquakes "may justly be regarded as the token of an incensed Deity," but the smaller shocks were viewed as a benefit to man and the earth.

Winthrop's views on the nature and origin of earthquakes were accepted for many years. In 1785, for example, Samuel Williams, another Harvard professor, presented "Conjectures on the causes of earthquakes," which appears to be no more than a paraphrase of Winthrop's *Lecture*. In 1793 Benjamin Franklin proposed only a slight variation on Winthrop's mechanism by favoring waves in a dense fluid, rather than expanding vapors, as the substance in the subterranean chambers (Paper 14). In 1813, Thomas Cooper returned to the expanding-gas concept in his essay on the earth's interior (Paper 15).

The true origin of earthquakes was to remain a matter for conjecture until many years later, but Winthrop's observations on the wave-like propagation of the earth have withstood the test of time.

REFERENCE

Hindle, B. 1956. *The Pursuit of Science in Revolutionary America*. Chapel Hill, Univ. of North Carolina Press.

BIBLIOGRAPHY

Alden, T. 1804. On Earthquakes. *Massachusetts His. Soc. Coll.* **9**:232-234.

Anonymous. 1757. The Philosophy of Earthquakes. *Am. Mag. and Monthly Chronicle for the American Colonies* **1**:23-24.

Barton, B. S. 1804. Memorandums Concerning the Earthquakes of North America. *Phila. Med. Phys. Jour.* **1**:60-67.

Doolittle, T. 1693. *Earthquakes Explained and Practically Improved*. Boston: B. Harris, 56p.

Dudley, P. 1789. An Account of the Earthquakes which Have Happened in New England. *Am. Mus. or Universal Mag.* **5**:363-365, 595-597.

Griswold, Samuel. 1813. Information Concerning the Earthquakes which Have Prevailed in the United States Since December, 1811. *Med. Repos.* **16**:304-309.

Mather, Increase. 1706. *A Discourse Concerning Earthquakes. Occasioned by the Earthquakes which Were in New-England, . . . June 16 and in Connecticot [sic] Colony, June 22, 1705*. Boston: T. Green, 131p.

Mitchill, S. L. 1815. A Detailed Narrative of the Earthquakes which Occurred on the 16th Day of December, 1811, and Agitated the Parts of North America that Lie Between the Atlantic Ocean and Louisiana. *Lit. Philos. Soc. of N.Y. Trans.* **1**:281-307.

Williams, S. 1785. Observations and Conjectures on the Earthquakes of New England. *Am. Acad. Arts and Sci. Trans.* **1**:260-311.

Winthrop, J. 1756. *A Letter to the Publisher of the Boston Gazette*. Boston, Edes and Gill, 7p.

1

Reprinted from pages i, iii, v–vi and 1–20 of *Earthquakes, the Works of God and Tokens of His Just Displeasure*, Boston: D. Henchman, 1727, 48p.

Mr. *Prince*'s
Fast & Thanksgiving
SERMONS
On the
EARTHQUAKE.

EARTHQUAKES *the Works of GOD & Tokens of his just Displeasure.*

Two SERMONS
On
PSAL. xviii. 7.
At the Particular *Fast* in *Boston*, Nov. 2.
and the General *Thanksgiving*, Nov. 9.
Occasioned
By the late dreadful

EARTHQUAKE.

Wherein

Among other things is offered a brief Account of the *Natural Causes* of these Operations in the Hands of God: With a Relation of some *late terrible Ones* in *other Parts* of the World, as well as those that have been perceived in *New-England* since it's Settlement by *English* Inhabitants.

By Thomas Prince, M. A.
And one of the Pastors of the South Church *in* Boston.

The Second Edition Corrected.

Psal. cix. 2. *The Works of the LORD are Great, sought out of all them that have Pleasure therein.*
Psal. lvi. 5. *Come & see the Works of GOD! He is terrible in his Doing toward the Children of Men.*

BOSTON in New-England:
Printed for D. Henchman, over against the Brick Meeting House in Cornhill.
MDCCXXVII.

The PREFACE.

Giving a Summary Account of the OC-CASION of the following Sermons.

ON the NIGHT after the Lord's Day Octob. 29. about 40 Minutes past X, in a calm & serene Hour, the Town of BOSTON was on a sudden extreamly surpriz'd with the most violent Shock of an EARTHQUAKE that has been known among us. It came on with a loud hollow Noise like the Roaring of a Great fired Chimney, but incomparably more fierce & terrible. In about half a Minute the Earth began to heave and tremble: The Shock increasing, rose to the Hight in about a Minute more, when the Moveables, Doors, Windows, Walls, especially in the upper Chambers, made a very fearful Clattering, and the Houses rock'd & crackl'd, as if they were all dissolving and falling to pieces. The People asleep were awakened with the greatest astonishment: many others affrighted run into the Streets for Safety. But the Shaking quickly abated, and in another half Minute intirely ceased.

THE Noise & Shakes seem'd to come from the Norwestward, and to go off Southeasterly; and so the Houses seemed to reel. Some Damage was done to the more brittle sort of Moveables, and some Bricks on the Tops of some Chimneys fell; but not an House was broken, nor a Creature hurt. At several times till Day-light, were heard some distant Rumblings, and some fainter Shocks were felt: But since, the Earth has been quiet in Town, tho' the minds of many continue very greatly & justly affected.

The PREFACE.

IN the Morning *between* X & XI, *a very full Assembly met at the* North Church *for the proper Exercises on so extraordinary Occasion. At* V *in the Evening, two very crowded Congregations assembled at the* Old & South Churches *on the same Account. At the* LIEUT. GOVERNOUR'S *Motion, the* Thursday *of the same Week was kept as a Day of extraordinary* Fasting & Prayer *in all the Churches in* TOWN : *When the First of the following* Sermons *was Preached. And the* Thursday *after, having been sometime before appointed by Authority for the General Yearly* Thanksgiving *thro' the Province; I tho't it not amiss to proceed on the Subject: That being touched afresh with the sense of our late eminent Distress & Deliverance, we might be raised up to the higher and livelier Acts of Gratitude & Praise.*

THUS of this awful Event in the Town : *But the Circumstance of it in other Parts I must leave to be observed in an* APPENDIX.

AND as for the SERMONS, *I have only to say --- That being 'penned in an unavoidable Hurry on such an Occasion! I must desire the curious Reader to make proper Allowances. I cou'd not easily mend them, at least to my own Satisfaction, unless their whole Frame & Expression had been almost intirely Changed; and this would have made them quite others than what I deliver'd, and have been too much imposing on those whose prevailing Desires were to see by the Press what they heard from the Pulpit. In hopes of their helping the Readers to some happy Improvement of this awakening Providence; I humbly commit them to the effectual Influence of the* SPIRIT OF GRACE, *the absolute need of which I desire to acknowledge in all my Labours.*

Boston, *Nov.* 20.   **T. Prince.**
1 7 2 7

GOD Shakes the Earth Because He is Wroth.

SERMON I.

Thursday Morning, *Nov.* 2.

PSALM XVIII. 7.

THEN the Earth shook and trembled; the Foundations also of the Hills moved, and were shaken, because He was Wroth.

I SHALL make no Reflections on the Occasion of these Words as they lie in the Psalm. I only propose to consider them as they are a meer Allusion to a most awful and astonishing Work of GOD, which we were all the amazed Witnesses of a few Evenings ago. *Then the Earth shook and trembled, the Foundations also of the Hills moved and were shaken, because He was Wroth.*

NOR shall I make any critical Remark on the Text, unless this one --- That in the Parallel Place of 2 Sam. XXII. 8. *The Foundations of the Hills* are called, *The Foundations of Heaven.* No doubt for the same Reason that the *Hills* are also called the *Pillars of Heaven*, when they are said to *tremble*, in *Job* XXVI. 11. And this because the Tops of the
Mountains

Mountains seem to touch the Clouds, and the Heavens themselves to bear upon them (j). In this respect the Foundations of the Hills appear to the Eye, with that Magnificence & Grandeur, as if they were the Pillars and Foundations of Heaven it self.

But tho' the words be a mere Allusion; yet there must needs be a Justice & Propriety in it: and they must allude to Realities: And what can they be but these,—
1. That when the Earth shakes & trembles, and the Foundations of the Hills move and are shaken, it is the great GOD himself that shakes and sets them a moving.
2. That He does this terrible Work because He is Wroth, and thereby He expresses his fearful Displeasure.

These two things are fairly implied in the Text, as the very Ground and Occasion of this proper Allusion. It is the most High GOD himself that as the First Cause makes the Earth to tremble: And this He does because He is Angry, and thereby He clearly shows that He is so. These may be therefore the Observations before us,

Obs. 1. *THAT when the Earth shakes and trembles and the Foundations also of the Hills move and are shaken, It is the Great GOD Himself that shakes and sets them a moving.*

The Psalmist here uses a Variety of Phrases expressing nearly the same Idea; to represent the Repetition and Fearfulness of this Divine Operation, and raise our Minds to the most solemn and sensible Conceptions of it. But for the Illustration of this, I shall endeavour these three things,
1. To give some Description of this dreadful Event of Providence.
2. To consider something of the natural Causes.

(j) *Jackson* on 2 *Sam*. XXII. 8.

AND

And then,

3. And lastly, LEAD you up to GOD as having the highest and principal Agency in this tremendous Action.

I. *TO give some Description of this terrible Event of Providence*, the shaking and trembling of the Earth.

AN Earthquake may very possibly be UNIVERSAL. There are doubtless Provisions enough in Nature to produce such a one. Every Place on Earth, at ev'ry Season of the Year, and at ev'ry moment of the Day and Night, is doubtless liable to these Concussions. Whole Continents have trembled at once, with their neighbouring Islands, tho' the deep Channels of the Seas lay between them. And as the same Causes may concur & operate in the four Quarters of the Globe at once, they may at the same time effect a universal Convulsion.

SUCH a Shake of the whole Earth there seems to have been both at the Passion and Resurrection of it's mighty REDEEMER. Matth. xxvii. 50,---52. *Jesus, when He had cried again with a loud voice, yielded up the Ghost: And behold the Vail of the Temple was rent in twain from the Top to the Bottom: and the Earth did Quake, and the Rocks rent,* &c. The Sun & Heavens had been vail'd with Darkness for three Hours before, at his dying Groans * : And now the Earth was struck with Trembling at his Expiration. And at his Rising out of the Grave, we read in the following Chapter at the first Verse, *And behold there was a great Earthquake.*

How fit was it that the whole Earth shou'd tremble; when the Maker, the Upholder, the Lord and the Saviour of it, was struggling in the Pangs of Death, and his Soul & Body were violently rending asunder? And why shou'd not the whole Earth be moved; when He break out of the Bowels of it, in

---

* Mark XV. 33,---37. Luk. XXIII. 44,---46.

which He had been imprisoned, to rescue it out of the Hands of those infernal Powers that had for so many Ages usurped the Dominion over it.

OF this surprizing Time it was foretold in, Hag. II. 6, 7. *For thus saith the Lord of Hosts, yet once it is a little while and I will shake the Heavens, and the Earth, and the Sea, and the dry Land : and I will shake all Nations* ; † *And the Desire of all Nations shall come, and I will fill this House with Glory, saith the Lord of Hosts.* And thus the Apostle explains it in *Heb.* XII. 26. -- e : tho' doubtless with a further View to a more astonishing Fulfilment at our Blessed SAVIOUR's Second Appearance.

HOWEVER, The Great and dreadful Day is hastening, when there shall be an universal Convulsion in this lower World. Of this amazing Time we read in Isa. XXIV. 1, &c. *Behold the Lord makes the Earth empty, and makes it waste, and turneth it upside down, and scattereth abroad the Inhabitants thereof.* --- *Fear, and the Pit, and the Snare, are upon thee, O Inhabitant of the Earth ! And it shall come to pass that He who fleeth from the Noise of the Fear shall fall into the Pit : And he that cometh up out of the midst of the Pit shall be taken in the Snare : For the Windows from on High are open, and the Foundations of the Earth do shake : The Earth is utterly broken down, the Earth is clean dissolved, the Earth is moved exceedingly : The Earth shall reel to & fro like a Drunkard, and shall be removed like a Cottage, and the Transgression thereof shall be heavy upon it, and it shall fall and not rise again.*

WHOEVER diligently reads this Chapter and attentively Compares it with 2 *Pet.* III. *Rev.* XX. and other Places of the Holy Prophesies, where the very same things are mentioned, will doubtless clearly see that these Predictions of an Universal Earthquake

---

† So, Ver. 21, 22,

refer

refer to the End of this present State of the World, when there will also immediately come on an universal Conflagration. First, The Earth will extreamly shake and reel to and fro: and therewith the Buildings on it must need make a dreadful crackling Noise and tumble to Ruins. On this account 'tis said, The Inhabitants shall *Fly from the Noise of the Fear* --- *i. e.* The fearful Noise of their Dwellings rending to pieces: But flying from this, they shall *Fall into the Pit.* --- *i. e.* The gaping Caverns of the Earth opening round about them: and those who escape from these shall be *taken in the Snare* --- *i. e.* Such an inextricable one as we have described in Psal. xi. 6. *Upon the Wicked He shall rain Snares, Fire and Brimstone, and an horrible Tempest: This shall be the Portion of their Cup.* The nature of this frightfull Noise, with the Pit and Snare, is therefore plainly pointed at in the words immediately following, wherein the Two general Causes are thus assigned --- *For the Windows from on high are opened,* * *and the Foundations of the Earth do shake.*

Of the same dreadful Time we also read in Rev. XVI. 15. --- *e* ‡ *Behold I come as as a Thief: Blessed is He that Watcheth* --- *And there were Voices and Thunders & Lightnings, and there was a great Earthquake, such as was not since Men were upon the Earth, so mighty an Earthquake, and so great: And the great City was divided into three Parts, and the Cities of the Nations fell: and great Babylon came into Remembrance before God, to give unto Her the Cup of the Wine of the Fierceness of his Wrath: And ev'ry Island fled away: and the Mountains were not found: and there fell upon Men a great Hail out of Heaven,* &c.

---

* A plain Allusion to the Lord's Raining down Fire and Brimstone out of Heaven on Sodom; and agrees with the miserable Desolation of the Earth by Fire, which we have described in the 4th, 5th & 6th Verses of the same Chapter.
‡ So *Rev.* VI. 12. --- *e.* VIII. 5. & XI. 13. --- *e.*

At the End of this present State of the World, will be such a universal sudden and destroying Earthquake. But till that amazing time comes on, our Blessed Saviour tells us; *There shall be Famines, Pestilences, Troubles, Great Earthquakes in diverse Places, and fearful Sights and great Signs from Heaven: That these are the Beginning of Sorrows, and the Signs of his Coming, and of the End of the World* ‖.

They are these PARTICULAR Earthquakes that we are now considering: of which there have been great Numbers of tremendous Instances in these latter Ages; and in which there has been a very great Variety.

They sometimes come upon some threatning and foreboding Symptoms: But for the most part at once, without the least Intimation or Suspicion. Sometimes in dreadful Tempests; but oftener in the calmest and serenest Seasons. They are sometimes of a narrower, at others of a larger Compass. Sometimes their Shocks are of a shorter, at others of a longer Duration: Some are of a lesser, others of a greater Violence: Sometimes they quickly cease and leave the Earth at Rest; at other times they soon repeat their terrible Efforts, and throw the Earth into new Convulsions.

Sometimes they make the Earth to jarr & vibrate, and sometimes to waver like the Waves of the Sea. In such a manner *was the Place shaken* where the *Apostles* assembled. *Acts* 4. 31. And thus *the Foundations of the Prison* where *Paul & Silas* lay *bound, were shaken* by the Great & sudden Earthquake, in *Acts* 16. 26. The Original Words in both those Places signify, to fluctuate like the Waves of the Ocean, or to reel, or rise and lower like the Ships upon them. (z)

‖ *Mat.* XXIV. 3,--8. *Mar.* XIII. 4,--8. *Luk.* XXI. 7---11.
(z) *Zanch. Bez.* &c. in *Leigh Crit. Sac.*

SOMA

SOMETIMES they alter the Channels of the Seas themselves, and fill up the most commodious Harbours. Sometimes they throw down Houses, or roll the Mountains upon them, and bury their Inhabitants under their Ruins: and thus their Beds are made their Graves, and their Dwellings turned into Tombs to cover them. Sometimes they open the Ground, and swallow up whole Streets & Cities into the Depths of the Earth in a moment: The Parents and their Children go down alive & shrieking, with all their Substance into the Pit together: They are covered up for ever: And either the Earth closes over them, or the gushing Floods of Waters in a moment overwhelm & drown them, and leave no sign of the Places were they have lived & sinned together.

SOMETIMES with the opening Earth there issues out a firy stream of Brimstone, that runs with an amazing Swiftness, devours every thing before it, and that none can escape. And sometimes they blow forth mineral and malignant Streams that destroy the People round about with raging Pestilence. Thus it was with *Korah* and his guilty Company. Numb. XVI. 31 --- 35. * *The Ground clave asunder that was under them, and the Earth opened her mouth & swallow'd them up, & their Houses, and all the Men that appertained to Korah, & all their Goods: They & all that appertained to them, went down alive into the Pit, and the Earth closed upon them, and they perished from among the Congregation. And all* Israel *that were round about them, fled at the Cry of them: for they said, Least the Earth swallow us up also: And there came out a Fire from the Lord, and consumed* Two Hundred and Fifty *more that were chief in the Wickedness.* And with this dreadful Eruption there immediately follow'd a *Plague*, that quickly destroy'd near Fifteen Thousand People.

* Numb. XXVI. 10. Psal. CVI. 17, 18.

AND

AND now, surely I need not describe the Surprize, the Anguish and Horror that seizes and distresses our Minds, when we are feeling our Houses rocking over our Heads and straining every Joint, and the Foundations of the Earth a moving under our Feet. *When the Centurion and they that were with him saw the Earthquake* at the Crucifixion of CHRIST, *They Feared greatly* †. And of the Terror of such an awful Event we read, 1 Sam. XIV. 15. *And there was Trembling in the Host, in the Field, and among all the People: The Garrison and the Spoilers they also trembled, and the Earth Quaked: So it was a very great Trembling*; or *a Trembling of GOD*, as it is in the Hebrew.

WE have oftentimes heard of such things at a Distance; and it's likely with a slighty Resentment, as remote or foreign Events that did not much concern or threaten us. But now we have known by Experience, something of the Dreadfulness of them.

WHILE our *Curtains* like those of the Land of *Midian* were trembling ‡--- Fear came upon us and Trembling also, which made the multitudes of our Bones to shake. Our Hearts were sore pained within us, and the Terrors of Death were fallen upon us. Fearfulness and Trembling seized us, and Horrour overwhelmed and cover'd us: And we were ready to fly as they fled from the Earthquake in the Days of *Uzziah*: ‖ Or to wish --- O that we had Wings like a Dove, that we might fly away and be safe from Destruction! Did we not see every one with their Hands on their Loyns, as a Woman in Travel, and all Faces turn'd into Paleness? *Alas! For* THAT NIGHT! it was Great, so that none was like it! It was even the Time of *Jacob*'s Trouble: Tho' we are saved out of it. Even when we remember, we are still afraid, and Trembling takes hold on our Flesh.

---

† *Matth.* XXVIII. 54. ‡ *Hab.* III. 7. ‖ *Zech.* XIV. 5.

And thus much of the general Appearance, Effects and Attendants of this awful Event.

We go on to consider,

II. *SOMETHING of the Natural, Instrumental or Secondary Causes of it.*

And here I shall be as brief as possible, least I keep you too long from the view of the *First and Principal Cause* in these great Affairs: who for ever has all created Things in his Hands, and always Acts on them in the wisest manner to fulfil his Designs. Nor shou'd I treat of these natural Causes, were we not to consider them as the very Works or Operations of God, and the Means he uses to accomplish his Pleasure. With a reverend Mind, and not a mere natural but divine Philosophy, let us search a little into them, and see --- How the mighty God invisibly works by sensible Causes, and even by those that are extreamly little and weak, produces the greatest and most terrible Effects in the World.

And here, I shall intirely omit the many fine but uncertain *Hypotheses* of the projecting Sort of Philosophers both ancient (s) and modern; such as of a central *Concave of Fire*, a vast internal *Abyss of Waters*, &c. I shall only take notice of *those Natural Causes, whose Operations are known to us all,* and conceived with Ease when applied to these Matters.

And I must in the first place observe, That in the very Constitution and Frame of this Earth, God has created and form'd it in such a particular manner as to prepare and dispose it for such fearful Convulsions.

He has not only made it of a very loose Contexture, thro' which the Air and Water and other subtil Materials can easily penetrate: But He has also formed it with a vast and inconceivable number of *Caverns* or

---

(s) See a large Collection of the ancient Opinions in SENACÆ Quest. Nat.

hollow

hollow Places within : (*v*) and in every Part He has either primarily lodged, or by the constant Circulation of Air and Water conveyed, great multitudes of *Sulphurious, nitrous, fiery* and *mineral Particles*, such as those in the Clouds, which are the natural Causes of Thunder and Lightning.

Now when these are put into Motion and strike against one another; we know with what amazing Violence they fly off from each other, as we see in our *Aurum Fulminans* and common *Gun-Powder*. So prodigious is their Expansion, that it is computed a Collection of them set on Fire, will in a moment spread so wide, as to possess above a *Million Times* the space they did before, while they lay in the Form of a compacted Body. They will first make their way into the easiest Passages, and if they have not a speedy Vent, agreable to their expansive and united Power, they will tare and rend away all before them.

And thus these diminutive Causes meeting and acting in a suitable Quantity, whether in the Air or Water of the hollow Vaults and empty Spaces below; they fly off from each other with the greatest Violence. They firstly issue or push every Vapour before them into the neighbouring Caverns: But still wanting Room, and especially if the Perspirations or Pores of the Earth are stopt either by Frost or Rains, whereby they cannot get thro' soon enough; they heave and raise the Earth, till they force their way into the Air above.

(*v*) Vid VAREN Geog. And the Observer has Himself been in many of these Caverns in England: Some of them crossing their Mines about 300 Foot below the Surface of the Earth: And others opening at the Bottom of Mountains & running deep under their Roots for a 1000 Yards inExtent; out of which others have branched into unknown and inaccessible Distances. So that in accounting for these natural Causes, He goes no further than He has seen with his Eyes.

Nos

Nor is this a Twentieth Part of our imminent Danger: There is a terrible *Atmosphere* over us, which few are aware of, that is found to press with the Weight of above *Two Thousand Pounds* * on every *square Foot* of the Surface of the Earth. And when by the violent Explosions made from beneath, the Vapours within are on a sudden discharg'd and leave a Vacuity; then this astonishing Weight, besides that of the Earth, immediately bares away every thing before it into the Spaces below, and sinks whole Cities & Mountains at once. Yea, tho' there shou'd be an Eruption thro' an easier Passage in any Part of the Land or Water about us, a great many Miles off; Yet Here, a whole Town being over such an emptied Cavern, it may sink in a moment: and even this without perceiving so much of the Shake of the Earth, as the Places nearer and round the Eruption.

Thus has the most High God, in the very Composition and Frame of this Earth, and by the Course and Laws whereby He governs all Nature, made sufficient Provision for such dire Convulsions. Thus He has placed us over great and hideous *Vaults*, large enough to receive the most spacious Cities, and ready to open when He sees it time to bury us in them: And there He lays his terrible *Mines*, to spring them at his sovereign Pleasure. And by such little weak Things, He causes great and marvellous Alterations in this lower World, and executes his most righteous Judgments.

I wou'd therefore proceed in the

III. and last place, *TO lead you up to GOD, as having the highest & principal Agency in this stupendeous Work.*

How exceeding apt are we to terminate our views and rest in second Causes; and look on them as certain fatal Things that move and act of themselves, and without Design or Reason. But this our Prospect

---

* Mr. *Boyl* found it to be 2592 Pounds.

is exceeding narrow, shallow, short & partial. And our Conceptions rising from it, are not only utterly wrong and contrary to the Truth of things, but are also opposite to the Being of Providence and the Nature of the Divine Perfections.

For this is to suppose, That when God had formed *material* Substances, He only made some *general Impressions* on them for general Ends, and then left them to Act according to those Impressions, without continuing his active or directive Influence on them: which is to make these material things now independent on Him in all their continued Actings, and is utterly inconsistent with his continual Providence. Or if He continues his Influence for those general Ends, that yet the most Wise God neither had nor has any particular View to all the Effects they were to produce in the World: which is intirely repugnant to his perfect Wisdom.

From the Operations of these natural *Agents*, we must therefore raise our Minds to their most wise Creator. We must look on *them* as his Workmanship, and upon Him as making and moving them in perfect Wisdom, and as having in View every Effect they were to produce when first He formed them and set them a moving. His all-wise and all-seeing Prospect must have taken and adjusted every Cause and Effect in the most beautiful Place in the Scheam of his Providence. Having therefore most wisely laid the universal Train in his Mind; He creates a vast Variety of Things as Means in which He pleases to Work, in the fittest Times & Places to fulfil his Designs. And having made his Instruments, He continually takes an effectual Care to impower and guide them to attain his Ends.

But however it is as to created spiritual and intellectual Agents; we know that merely *material* Substances can only move in the Circuits of Nature as they are moved by Him. We see they always and every where move and act in a most curious and regular manner with respect to each other, as if they per-

fectly

fectly knew each other's particular Circumstances, and were the most intelligent Agents. But they are uncapable of knowing the wonderful Laws by which they are Governed. They are therefore constantly Guided by the Wisdom and Power of Him that made them: He must continually hold them in his immediate Hands, and both impower and direct them in all their Actions. And what we call the *Laws of Nature*, are only the usual Methods in which He is pleased to Work in the World; and from which He sees not cause to depart, but in some extraordinary Cases where his usual manner of working cannot reach his Designs.

And in a particular manner ---- To what else can we in Reason assign that admirable Force and Action of those peculiar Substances that are the Causes of Earthquakes in *their violent flying off from each other*, but to the immediate Influence and Action of God? Those merely material, exceeding small and senseless Atoms, that lie intirely dead and unactive, are no sooner disingag'd from each other; but such an amazing Force is in a moment inspir'd into them from some other Agent, as to set the largest Countries of the Earth a Trembling with all their Mountains & Cities upon them, and to heave them up and rend them to Pieces.

If you were to see Two *mountainous Rocks* to lie perfectly still while they touched each other; but being by some Means removed to a small and particular Distance, then of themselves to fly off from each other with the greatest Violence --- wou'd you not ascribe it readily to some mighty and invisible Agent that with such a wonderful Force shou'd beat them away? Why, The smallest Atoms of Matter are of themselves as Dead and unable to fly off from each other as the greatest Rocks in the World.

And what mighty and invisible Agent shou'd this be but God? These diminutive Substances are ever dispersed all over the Earth, the Air and Seas: They rise with the finest Vapour into the cloudy Regions:

They distill in every Shower: They peirce into ev'ry Vacancy of the Earth below: They mingle with every Substance: And when their collected Parts are dis-join'd and remov'd to such particular Distances, they are always in a moment impres'd with the same astonishing Force and Violence, tho' they were absolutely still before. That mighty and invisible Cause which moves them is every where always Present, attending these numberless and most minute materials, and moving them all in the most regular and intelligent manner. He moves them only at such a determinate Distance, with such a uniform measure of Force and Activity, and to such a Degree of Expansion; and then having performed his Work, and gained his End, He drives them no further.

And who can this all-present, all-wise, all-powerfull and all-vigilant Being be, but the most Glorious GOD? It is He that holds all Nature in his sovereign Hands, and continually rules it in perfect Wisdom. It is He that Acts on these things He has made for such wonderful Ends, and that guides and imploys them as Means to produce such dreadful Concussions as Earthquakes in this World below.

So *Job* that Divine Philospher, discovers & owns of the Invisible GOD, in Chap. ix.4, &c. *He is wise in Heart & mighty in Strength --- which removeth the Mountains, and they know not: which overturneth them in his Anger: which shaketh the Earth out of her Place, and the Pillars thereof tremble.*

So King *David* in Psal: lx.2, 3. *Thou hast made the Earth to tremble, Thou hast broken it: Heal the Breaches thereof, for it shaketh: Thou hast shewed thy People hard things, Thou hast made them to drink the Wine of Astonishment.* And agreable to this are those lofty Expressions in Psal. civ.32. *He looketh on the Earth, and it trembleth: He toucheth the Hills and they smoke.*

AND

And thus we see the Great God has the chiefest Agency in these wonderous Works. Let us apply ourselves now to Improve this Doctrine.

And here,

1. *IN these affrightful Shakings of the Earth we see the dreadful Majesty of the Glorious GOD.*

What more awful or lively Instance can there be of this, than to see the Earth & Mountains trembling at his Presence? Thus the ancient Sinai did; of which it is said, That *Because the Lord descended on it, the whole Mountain quaked greatly.* * And when He went out of *Seir*, when He marched out of the Field of *Edom*; *The Earth trembled from before the God of Israel,* † *The Earth shook at the Presence of GOD, even Sinai it self moved at the Presence of the God of* Israel. ‡

With God is most terrible Majesty: And when He has a mind to show it, He can easily and in a moment do it in such an astonishing manner as to affright the hardiest Creature. He can put all Nature, even the Great & inanimate Parts of the World into such a Commotion, as to make us see in a most sensible manner, the terrifying Actings of his Powerful Presence, and excite the highest and most awful Reverence of Him. He can make the heavy & dull Earth to tremble, as if it felt the Force of those awakened Passions that shou'd rise in our Minds at the Appearance of God, and as if it were mov'd with the Fear of it's present Destruction. The Everlasting Mountains are scattered, and the perpetual Hills bow down before Him. ‖

2. *WHO will not then fear & tremble before this Great & dreadful Being?*

He that can shake the Earth out of Her Place in a moment without any Instruments; for He needs them not, nor is confined to them. Or if He sees fit to use them, He has always a sufficiency of them at Hand to imploy as He pleases. He can improve

**Exo.* XIX 18. †*Judg.* V. 4, 5. ‡*Ps.* LXVII 7, 8. ‖ *Hab.* III. 6.

the minutest of Atoms to shatter the Earth: And when He is exerting his Power upon them in the invisible Caverns below; who can tell *how far* He may do it in a greater measure, to the speedy & utter Destruction of those that neither fear or regard Him? He that exerts such a Degree of Power as He does in and by them, may as easily put forth a Million times Greater, and in a moment blow up the Earth, with all things on it, and disperse it into a Vapour.

HEAR now this O foolish People and without Understanding! which have Eyes and see not; which have Ears & hear not! Fear ye not me, saith the Lord? will ye not tremble at my Presence? *. It is the utmost Degree of Stupidity, to be fearless of Him; and a most provoking Offence to his infinite Majesty. Such fool-hardy Sinners He therefore admonishes in those awful Terms. † *Now, Consider this Ye that forget God! least I tare you in pieces, and there be none to deliver.*

LET us then have a Care of Returning to our former Security: And let us bare in our Minds a lively Sense of our continual Danger. Let our Flesh still tremble for Fear of GOD; and let us be ever afraid of His Judgments. Let us stand in the greatest aw of this most Glorious Being, and not Sin against Him. He is always Present, and the same Holy, Mighty and Terrible, as He appears to be in the most hideous Earthquake.

LIKE the astonished *Saul*, while he *trembled* and fell down to the Earth: Let us now earnestly cry---- *Lord what wilt Thou have me to Do* ‡ Or as the awakened *Jailour*, at *the great & sudden Earthquake* wherein the Foundations of his Prison at *Philippi* were shaken; came in *Trembling* and fell down before the Apostles and cried --- *Sirs, What must I do to be saved?* ‖ In the affecting Remembrance of this awful Providence and in the lively view of our

---

* *Jer.* V. 21, 23. † *Psal.* L. 22. ‡ *Acts* IX. 4—6. ‖ *Acts* XVI 26---30.

continual

continual Danger; Let us constantly adore and mention the dreadful Name of GOD with Reverence: and as the Apostle exhorted those *Philippians*, Work out our Salvation with Fear and Trembling. * *When* Ephraim *spake trembling, He exalted Himself in* Israel; *But when He offended in* Baal, *He Died.* †

But alas! The most frightful Appearances will signify nothing to bring us to a due Awe of the Glorious God, unless His Spirit shall attend and speak in a sensible manner. 1 Kings xix. 11,----13. *And behold the* Lord *passed by, and a great & strong* Wind *rent the Mountains, and break in pieces the Rocks before the* Lord; *but the* Lord *was not in the Wind*: *And after the Wind an* Earthquake; *but the* Lord *was not in the Earthquake*: *And after the Earthquake a* Fire; *but the* Lord *was not in the Fire*: *And after the Fire a* Still small Voice: *And it was so*; *when* Elijah *heard that,* Then *he wrapt his Face in his Mantle*, &c. With these terrible Works, Let us earnestly look and pray for the Holy Spirit to follow and speak in an effectual manner: And that He may be Poured out as He was when the *Place was shaken* where the Primitive Christians assembled, *Acts* iv. 31.

3. *HEREIN we also clearly see the utter Uncertainty & Insufficiency of every thing on the Earth.*

We see how they all continually stand on dangerous and uncertain Ground. Their very Foundations may fail in a moment. In an Hour we think not of, they may sink down with us, who have placed all our Affection and Dependance on them; Go down together into the Gulphs below, and be buried up together, till the Heavens be no more.

And alas! How empty and vain did they look in *That Night*; while we were shivering with them over

---
* *Phil.* II 12. † *Hos.* XIII 1.

the

the Pit, and in equal Hazard of being crushed to pieces? cou'd they yield us then any Support or Solace? Did they then appear to be Precious? Or cou'd we find any Comfort in our having Pursued them with so much Eagerness, and valued them more than the Favour of GOD and an Interest in Him.

Alas! WE then clearly saw the Vanity of these Earthly Things; we saw thro' their deceitful Appearance; and then we also understood our Delusions. O! That now having gained such a Sight and Discovery, we might never lose it again; nor so highly value, nor so earnestly seek, nor so securely rely on these doubtful things, that may so suddenly fail us, and leave nothing to comfort us in the last Extremity.

4. *HERE likewise we see the great Folly and Misery of those that live unprepared for such a fearful Hour.*

WE have in some measure seen with Astonishment how unexpectedly such a moment may come: and we know not how soon we may be surprized with another, and ten times more terrible. Even while we are securely Sleeping, as many were then, we may be rouzed up in an Instant to see all things dissolving about us, or feel our selves descending into the Depths Below. But O the Dismay and Confusion that will then seize the Soul unprepared for such a dismal Scene.

THEN the most High, whom you have continually slighted, will take his turn to despise and neglect you. "Because He had called, and you had refused; He had stretched out his hand, and you regarded him not, but set at nought all his Counsel and wou'd none of his Reproof: He will laugh at your Calamity and mock when your Fear comes; when your Fear comes as Desolation, and your Destruction comes as a Whirlwind; when Distress and Anguish comes on you. Then shall you call on him, but He will not answer: you shall seek him earnestly, but you shall not find him. *Prov.* 1. 24,---28.

AND

And O vile and guilty Earth! The Time is approaching, when the glorious God shall appear to search out and disclose thy Wickedness, and to shake it out of Thee in the Day of his Anger. *Tremble Thou Earth at the Presence of GOD!* For "a Fire shall go before him, and shall burn up "his Enemies round about: His Lightenings shall "enlighten the World: And the Earth shall see "and tremble: The Hills shall melt like wax at "his Presence, at the Presence of the Lord of the "whole Earth: The Heavens shall declare his Righ-"teousness, and all the People shall see his Glory. * In vain then shall the Wicked "go into the Clefts "of the Rocks, and into the Tops of the ragged "Rocks, for Fear of the Lord, and for the Glory "of his Majesty, when He arises to shake terribly the "Earth. † But "*Zion* shall hear and be Glad, "and the Daughters of *Judah* Rejoice, because "of his Judgments ‡.

5. and Lastly, *HOW exceeding safe and happy are they who have the mighty GOD for their Friend and Redeemer?*

They that have received His Son, forsaken their Sins, and obeyed his Gospel. Their Offences against Him are Pardoned: He is become their constant Friend: He has engaged to be so by an Everlasting Covenant, that shall never be broken. In his Almighty Hands and under his Eye, they are always secure: And they need "not Fear, tho' the Earth "be removed, and tho' the Mountains be carried in-"to the midst of the Sea, tho' the Waters thereof "roar and be troubled, tho' the Mountains shake "with the swelling thereof: The Lord of Hosts is "with them, The God of Jacob will be their "Strength and Refuge, and a present Help in their "Trouble. *Psal.* XLVII. 1, --- 7.

---

* *Psal* XCVII. 3 --- 7. † *Isa* II. 20. 21 *Rev* VI. 12, --- e.
‡ *Psal.* XCVII. 8.

Such as these, our Blessed Saviour rather exhorts to rejoice, even in the midst of those Tumults and Confusions of Nature that shall usher Him into these visible Heavens. Luk. xxi. 25---28. *And there shall be Signs in the Sun and in the Moon, and in the Stars ; and upon the Earth Distress of Nations, with Perplexity, the Sea and the Waves roaring ; Mens Hearts failing them for Fear, and for looking after those things which are coming on the Earth: For the Powers of Heaven shall be shaken : And then shall they see the Son of Man coming in a Cloud with Power and great Glory. And when these things begin to come to pass ;* THEN LOOK UP AND LIFT UP YOUR HEADS ; *For your Redemption draws near.*

I shall CONCLUDE with those Words which are spoken of that Great and terrible Day, in Joel III. 14,---16. *Multitudes, Multitudes in the Valley of Decision ; For the Day of the* LORD *is near, in the Valley of Decision. The Sun and the Moon shall be darkened, and the Stars shall withdraw their shining: The* LORD *also shall roar out of* Zion, *and utter his Voice from* Jerusalem : *And the Heavens and the Earth shall shake:* BUT THE LORD WILL BE THE HOPE OF HIS PEOPLE, AND THE STRENGTH OF THE CHILDREN OF Israel.

The END of the First *SERMON.*

[*Editor's Note:* This pamphlet was originally published with two sermons and an appended description of the earthquake (pp. 21–48), which have been omitted here.]

2

Reprinted from pages i, iii and 5–31 of *A Lecture on Earthquakes*, Boston: Edes and Gill, 1755, 38p.

Mr. *Winthrop*'s
LECTURE
ON
EARTHQUAKES.

A LECTURE ON EARTHQUAKES;

Read in the Chapel of *Harvard-College* in *Cambridge*, N. E. November 26th 1755.

On Occasion of the great *EARTHQUAKE* which shook NEW-ENGLAND the Week before.

By *John Winthrop*, Esq;
Hollisian Professor of the Mathematics and Philosophy at *Cambridge*.

Published by the general Desire of that *Society*.

Subterraneous caverns and vulcanos, if well considered, will be found to be wise contrivances of the Creator, serving to great uses of the Globe, and ends of GOD's government. In all probability, these things may minister unto many secret, grand functions and operations of nature in the bowels of the earth.   Dr. DERHAM's Physico-Theol.

BOSTON; NEW-ENGLAND:
Printed and Sold by *Edes & Gill*, at their Printing-Office next to the Prison in *Queen-Street.* 1755.

A LECTURE
On EARTHQUAKES.

YOU may juftly expect, that the great EARTHQUAKE, which fo lately ᵃ fpread terror, and threatened defolation throughout *New-England*, fhould take me off from my ftated courfe of lectures, to inquire into the probable caufes of fo formidable a phænomenon. The fubject is curious, and at prefent engages the attention of many perfons; and the difcuffion of it may help to extend your views.

An Earthquake, you all know, is an agitation or fhaking of fome *confiderable* part of the earth, and that by *natural* caufes; in contradiftinction to the fhaking of a *fmall* part of it by *artificial* methods. The degrees of this fhaking are very various;— from the fmall jarrings, which are but juft perceptible, to thofe violent fuccuffions, which have altered the face of whole countries.

ᵃ This great Earthquake happened on *Tuefday* the 18th of *November*, 1755, about a quarter after four in the morning. There was another much fmaller fhock an hour and quarter after this, viz. at 5ʰ 29′; and a third, on the *Saturday* evening following, at 27′ after 8. Since the reading of this lecture, there has been another fmall fhock, viz. on *Friday* the 19th of *December* in the evening, exactly at 10 o'clock; the fky being then perfectly clear, and a very gentle gale at S. W. It was preceded by the peculiar noife of an Earthquake about 3 or 4 feconds, and the jarring lafted near as long; caufing the window-fhutters and door of the chamber, in which I then was, to clatter. Thofe of my family, who were in a lower room, perceived nothing of the fhake, though they heard the noife. Thefe are the only fhocks that I have been fenfible of; though it is faid, that many others have been felt in the Province of *New-Hampfhire*, fince the firft great one.

tries. These shakings are for the most part (I believe, always) præceded or attended by an hollow rumbling noise, something like what is called *heavy thunder*; which is usually greater or less, according to the degree of the shake. Naturalists have distinguished earthquakes into two kinds; one, when the motion is horizontal, or from side to side; the other, when it is perpendicular, or right up and down. This distinction may, for what I know, be just; and yet perhaps earthquakes more commonly consist in a kind of *undulatory* motion, which may include both the others. For as a wave of water, when raised to it's greatest height, subsides, and in subsiding spreads itself horizontally; so in like manner, a *wave of earth*, if I may be allowed the expression, must in it's descent partake both of an horizontal and perpendicular motion at the same time. And for the same reason, it must have had both these motions in it's ascent; but those particles, which had been carried forward in one direction in the ascent, will return in a contrary direction in the descent. This has been evidently found to be the case in the more violent earthquakes; and probably the reason why it has not been universally found so, was, the difficulty of distinguishing these two motions from one another, when each of them has been but small. Though the ancient *Ægyptians* and *Chaldæans* are said to have been able to foretell earthquakes, yet it is very certain, from all the accounts we have, that these agitations of the earth do no where observe any order or regular period in their returns; but at sometimes, recur more frequently; at others, after longer intermissions. If therefore they pretended to foretell them at all, they must have done it, not from any knowledge they had of their nature and causes; but only by the vain arts of judicial astrology;—a kind of learning, it seems, which, futile as it is, was held in high repute among them. No countries, of which we have any knowledge, are exempt from these agitations; but some are more subject to them than others; and it is observable, that those which abound most with combustible minerals, as fossile coals, sulphur, nitre, &c. are the most exposed to them. Many of these countries, too, have certain mountains called *vulcanos*, which are almost perpetually burning, and throwing out fire,

fire, and smoke, and ashes; their entrails probably consisting chiefly of such sort of minerals. It is observed, however, that about the time of an earthquake in those places, these vulcanos rage more furiously, projecting stones and cinders to a great height in the air; and pouring out whole rivers of liquid fire, which carry such a devastation, wherever they run, as no human art can either prevent or repair. Several such there are in the *Molucca* islands in the *East-Indies*, almost under the equinoctial; and *Iceland*, under the polar circle, has four or five, besides the noted *Hecla*. *Vesuvio* near *Naples* is very remarkable; but there is none more famous than that in *Sicily*, now known there by the name of *Monte Gibello*, as it was formerly by that of *Ætna*; whose eruptions Virgil has described in so picturesque a manner, that I cannot forbear repeating a few lines from him.

———Horrificis juxta tonat Ætna ruinis:
Interdumque atram prorumpit ad æthera nubem
Turbine fumantem piceo et candente favilla;
Attollitque globos flammarum, et sidera lambit.
Interdum scopulos avulsaque viscera montis
Erigit eructans, liquefactaque saxa sub auras
Cum gemitu glomerat, fundoque exæstuat imo.
Æneid. III. 571. seqq.

This description [b] of these phænomena is perhaps not exceeded by any extant; except by those passages in the holy scriptures which ascribe these effects, as true philosophy does all those
which

---

[b] For the sake of English readers, Dr. Trapp's translation of this passage is here inserted.

'———— *Ætna* thunders nigh
' In dreadful ruins. With a whirlwind's force
' Sometimes it throws to heav'n a pitchy cloud,
' Redden'd with cinders, and involv'd in smoke;
' And tosses balls of flame, and licks the stars.
' Sometimes with loud explosion high it hurls
' Vast rocks, and entrails from the mountain torn;
' With roaring noise flings molten stones in air,
' And boils, and bellows, from it's lowest caves.

which we call *natural* effects, to the agency of GOD. It is HE, ' who removeth the mountains, and they know not; who ' overturneth them in his anger; who shaketh the earth out of ' her place, and the pillars thereof tremble.' ' HE looketh on ' the earth, and it trembleth; HE toucheth the hills, and they ' smoke.' ' The pillars of heaven tremble, and are astonished ' at HIS reproof.' Nothing can equal the sublimity, the grandeur of these images. But to proceed.

That the earth has been, in all ages, and in most parts of it, subject to these agitations, history affords us but too many proofs. It is not, however, my design to enter into a detail of the dismal events of this sort, left upon record. I shall only extract from authentic accounts a few of the most striking particulars, in order to give you some idea of the dire effects, which such convulsions of the earth are capable of producing.

Imagine then the earth trembling with a huge thundering noise, or heaving and swelling like a rolling sea:—now gaping in chasms of various sizes, and then immediately closing again; either swallowing up the unhappy persons who chanced to be over them, or crushing them to death by the middle:—from some, spouting up prodigious quantities of water to a vast height, or belching out hot, offensive and suffocating exhalations; while others are streaming with torrents of melted minerals: —some houses moving out of their places; others cracking and tumbling into heaps of rubbish; and others again, not barely by whole streets, but by whole cities at a time, sinking downright to a great depth in the earth, or under water:—on the shore, the sea roaring and rising in billows; or else retiring to a great distance from the land, and then violently returning like a flood to overwhelm it; vessels driven from their anchors; some overset and lost, others thrown up on the land:—in one place, vast rocks flung down from mountains, and choking up rivers, which, being then forced to find themselves new chanels, sweep away such trees, houses, &c. as had escaped the fury of the shock; in another, mountains themselves sinking in a moment, and their places possess'd by pools of water:—some people running about pale, with fear, trembling for the event, and ignorant
whither

whither to fly for shelter; others thrown with violence down on the ground, not being able to keep on their feet; and others shrieking or groaning in the agonies of death :—even the brute creation manifesting all the signs of consternation and astonishment :—Imagine these things to yourselves, and you will then have a view, though but an imperfect one, of some of those images of horror and desolation, which accompany the more violent earthquakes.

But I will dwell no longer on these tragical scenes. Those of you, who are desirous of farther information, may meet with it in the Philosophical Transactions;[c] particularly in the accounts of those horrible earthquakes, which almost desolated the islands of *Jamaica* and *Sicily*, 63 years ago; in the latter of which it was computed, that about 60,000 persons perished; which was very near one quarter of the whole number of inhabitants. To relieve you and my self under such melancholy prospects, I will turn my thoughts to the theory of these most formidable phænomena, as soon as I have made two or three observations on the earthquakes we lately felt.

GOD be thanked, all earthquakes are not formidable in so high a degree. Those of *New-England*, in particular, which have indeed greatly and justly alarm'd the inhabitants, have never destroyed them. For tho' this country, we know, has been visited with earthquakes from it's first settlement by the *English*; yet, so far as my information reaches, not a single life has been lost by any of them; and perhaps never so much damage done to our buildings as by the last great shock: As to which I would observe,

First, That it certainly began with an *undulation* of the earth, as I have been assured by some who were then awake; tho' I think it would not have been easily concluded from the effects, that the earth had had any other than an horizontal motion; those effects which were generally taken notice of, being chiefly, if not only, such, as a perpendicular motion would not, but an horizontal one would, have produced. Such, for instance, were the dashing of liquors over the sides of open vessels; the over-

B setting

[c] *Lowthorp*'s Abridgement, Vol. II. p. 400—419.

setting many things in houses; and the throwing of bricks from off the tops of chimnies to some distance. In order to estimate the velocity with which some were thrown from my chimney, I measured the greatest distance on the ground to which any of them had reach'd, and found it to be 30 feet; and the height of the chimney from which they fell was 32 feet. Now bodies fall thro' 16 feet nearly in one second of time; and the times, in which they fall thro' other heights, are in the subduplicate ratio of those heights. From whence it follows, that the velocity, wherewith those bricks were thrown off, was that of above 21 feet in 1" of time. For the subduplicate ratio of 32 to 16 is the same as the simple ratio of 30 to a little more than 21. It will be impossible, I believe, ever to determine with exactness the real spaces thro' which any of our buildings vibrated, in this reciprocating motion of the earth. It may be observ'd, however, that the shorter these vibrations are supposed to have been, the quicker or more frequent they must have been; the number of them in a given time being reciprocally as the length of each. Thus, for example, if each vibration had been of one foot in length, then there were 21 of them in 1" of time; but if each were of 6 inches, then there were 42 of them in a second; and so on. Possibly, some of these reciprocations might be as quick as those of a musical chord.

But it is not to be doubted, that the velocity, wherewith our buildings were agitated, was different in different places. It was different also, as I apprehend, at different heights. This I collect from the observation, that a key, which was thrown off a shelf in my house, was not thrown so far, in proportion to the height thro' which it fell, as the bricks were from the top of the chimney. Hence it appears, that our buildings were *rocked* with a kind of angular motion, like that of a cradle; the upper parts of them moving swifter, or thro' greater spaces in the same time, than the lower. ᵈ This perfectly agrees with the idea of an

undulatory

ᵈ That the vibrations of our buildings, and especially of the higher parts of them, were in fact extremely swift, appears from some effects, which had not come to my knowledge, when I wrote the above paragraphs. ' The ' bursting of a distiller's cistern, by the agitation of the liquor in it ; and the
' breaking

undulatory motion of the earth; as you may clearly conceive by turning your thoughts to the case of a vessel floating at rest upon stagnant water, and then suddenly agitated by a great wave rolling under it. In the motion of ascent, the mast of the vessel would be thrown forward, in the same direction as the wave was moving; and in the motion of descent, backward, or in the contrary direction; and in both these cases, the top of the mast would move thro' greater spaces than the bottom. As

' breaking off the spindle of the vane on *Faneuil-hall*, in *Boston*,' are arguments of it. This spindle, if I am rightly informed, was a pine stick, of about 5 inches in diameter, at the place where it was snap't off; and 10 feet in height; and the weight of the vane on the top of it was about 30 pounds. But nothing more strongly infers the excessive swiftness of these vibrations, than the bending of the wind-vanes on some high steeples. One at *Boston* was bent at it's spindle 3 or 4 points of the compass; and another at *Springfield*, distant about 80 miles in a right line westerly from *Boston*, was bent to a right angle; as I was assured by an eye-witness. This effect seems to be owing to a smart shock upon these spindles, in a direction perpendicular to that in which the vanes happened to stand; which in an instant gave them a very great velocity. The motion from below being given in the first place to the spindles, they were jerk'd forward so swiftly, that there was not time to have it communicated to the extreme parts of the vanes; which therefore were left behind, and remain'd in their former places. This may be illustrated by the following instance. ' When a door is half open, and ' moving very freely on it's hinges; if a pistol be fired against it, the ball ' will go through the door, without moving it out of it's place: because the ' motion of the ball is communicated but to a few parts of the door;' viz. only those which the ball drives before it. Dr. DESAGULIERS's Experimental Philosophy, Vol. I. p. 419. The resistence of the air might also contribute to this effect. The resistence, which a fluid makes to the motion of bodies in it, is, *cæteris paribus*, as the square of their velocity. When the motion, then, is not very swift, a fluid affords a pretty easy passage to bodies moving in it; but when the motion is so swift, that there is not time for the particles of the fluid to give way, and make room for the moving body, in this case a fluid will resist as much as a solid. An instance of this we have in the book now cited, p. 420. ' A musket being fired against ' water, the bullet was beaten to pieces upon the surface of the water. ' And thus in the present case, the velocity, attempted to be given by those spindles to their vanes, may be supposed so vastly great, that the air might resist their motion, in the manner of a firm, immoveable obstacle; which resistence therefore would be equivalent to the stroke of an hard body against them. In either way of accounting for this effect, it is visible, that the velocity, wherewith the tops of those steeples vibrated, must have been extremely great.

As it is certain, that in the great shock, the earth had an horizontal motion; so it appears with the most sensible evidence to me, that in the shock we felt the *Saturday* evening following, at 27′ after 8, there was a perpendicular motion of the earth. I was then sitting on a brick hearth, and felt the motion of the bricks distinctly under my feet. It was not a motion of the whole hearth together, either from side to side, or up and down; but of each brick separately by itself. Now as the bricks were contiguous, the only motion, which could be communicated to them separately, was in a perpendicular direction; and the sensation excited in me was exactly the same, as if some small solid body, by moving along under the hearth, had raised up the bricks successively, which immediately settled down again. The motion of the earth in this instance plainly appeared *undulatory* to me; and this shock, I apprehend, was occasioned by one small *wave of earth* rolling along, but not with a very swift motion. For the velocity of it's progress was considerably less than that of sound, which moves about 13 miles in a minute; as appeared from hence, that the roar of this earthquake might be heard at least half a minute before the shake was felt. Which also argues, that the shock began at some considerable distance from this place. The same remarks may be applied to the great shock; only this began with two at least, if not three *waves*, of much greater breadth and height. The latter part of this shock was *tremulous*, consisting chiefly in vibrations which succeeded one another with extreme quickness; and, as I take it, was owing to the efforts of the earth to recover the position, out of which it had been violently thrust, during the undulatory motion: Much in the same manner, as the reciprocations of a musical chord are occasioned by it's endeavours to restore it self to that situation, which it had before it was struck. As soon as the stroke, which bent it, ceases, the chord does not barely regain it's rectilinear figure, but bends itself almost as far the contrary way; and thus continues bending and unbending itself with great quickness, till it's motion is gradually destroyed, and at length it settles into a state of rest. Both these shocks, then, seem
evidently

evidently to have been of the *undulatory* kind; and to have differed in degree only. [e] I would obferve

Secondly, that the *duration* of the great fhock was *longer* than has been ufually obferved. If my memory fails me not, even thofe earthquakes, which have brought on the moft amazing cataftrophes, have commonly done their execution in one or two minutes; whereas this fhock with us lafted at leaft four; taking in the whole of the time, from the firft agitation of the earth, till it was become perfectly quiet; though the violence of the fhock did not laft above half fo long. This I am affured of, partly from the obfervations of fome Gentlemen, who were up, and looked on their watches, when it began and ended; and partly from my own, which were as follows. The preceding noon, I had adjufted both my clock and watch, by a meridian line; and the following noon I found that the watch had kept time exactly. Being awaked by the earthquake, I lay till the violence of it feemed to be over, for the fecond time; for it had a little abated before, as if it were going off, and then inftantly began again with redoubled fury. Till then I forbore to rife, becaufe the agitation was fo vehement, that I concluded it would be very difficult, if not impracticable, to go from the bed to the chimney, without being thrown down; and therefore thought it beft not to attempt it. The fpace of time, in which I lay awake, I cannot think to be much, if any thing, lefs than 2′. This was the conjecture I formed at that time; though it being but conjecture, I would not lay very great ftrefs upon it, were it not fupported by concurring obfervations. On the fecond abatement I rofe, and lighting a candle, looked on my watch, and found

[e] I take the laft fhock, on the 19th of *December*, to have been alfo *undulatory*. For it's being more fenfible above ftairs than below, as was obferved in Note [a], is an indication, that the motion of the houfe was of the *rocking* kind, and this implies *undulation* in the earth.

It may not be amifs to obferve here, that thefe undulations of the earth are not to be fuppofed to follow one another in fo orderly a manner, as the waves of the ocean often do: nor that they all move forward exactly in one direction. They rather refemble the agitations of a *tumbling* fea, whofe furges rife and fall irregularly, and follow different courfes in different places; and in the fame place too, at different inftants of time. This appears from that remarkable effect, of fome chimnies being partly turned round. One of mine looked as if it had been twifted, quite down to the roof.

found it to be 15′ after 4. The shock then was not quite over, but the windows continued rattling for about a minute longer, as near as I can remember; for the shock went off very gradually. As soon as I had looked on the watch, I went directly to the clock, which was in another chamber, that I might see whether that agreed with the watch; and found that it was stop'd at 4^h 11′ 35″. It's stopping, however, was not, immediately, owing to the violence of the shock, though several clocks, and watches too, at *Boston* are said to have been stop'd by it; but to the following accident. Having some time before used a pretty long glass tube, in a particular experiment, I had shut it up in the clock-case for security; and this tube, being overthrown by the earthquake, lodged against the pendulum, and stop'd it's motion. By this accident, the beginning of the earthquake, I conceive, is determined with all the exactness that can be desired; for, so far as I can learn, the first shake was violent enough to overset so tall, slender a body, and standing in a position so near a perpendicular, as that tube; and it was impossible for the pendulum to make one oscillation, after the tube had struck against it. Now from the time when the clock stop'd, to my looking on the watch, it was about $3'\frac{1}{2}$; and the jarring was not quite over, till about a minute after this: So that I think I speak within bounds, when I say, that this shock with us lasted at least 4′. In other places, it's duration might possibly be different. I observe

Thirdly, as to the *course* of this earthquake, that it seems to have been nearly *from N. W. to S. E.* I was informed a few minutes after the shock, by a person who was upon the *common* in this town at the time, that the noise began about the N. W, and came on from thence, and pass'd away toward the S.E; and other accounts, which I have since met with, agree with this. Those who were in such clear, open places could make the best judgement in this matter; for such as were within doors, or surrounded with buildings, might easily be misled by the various reflections of the sound. I am induced to give the greater credit to this information, by what I observed my self. For the key before spoken of, as thrown from off a shelf in my house, was found at a place on the floor, which bore very near N. W. of

the

on Earthquakes.

the place from which it fell; though the situation of it before it's fall was such, that it might have been thrown in several other directions as well as that, had the course of the earthquake been different. ᶠ

Let

ᶠ An account which we have lately received from the *West-Indies* agrees very well with the supposition, that our earthquake proceeded from hence southeastward. The account is, 'that on the 18th of *November*, about two o'clock in the afternoon, the sea withdrew from the harbour of St. *Martin*'s, leaving the vessels dry, and fish on the banks, where there us'd to be 3 or 4 fathom water; and continued out a considerable time, so that the people retired to the high land, fearing the consequence of it's return; and when it came in, it arose 6 feet higher than usual, so as to overflow the low lands. There was no shock felt at the above time.' As this extraordinary motion of the sea happened about 9 hours after we had the great shock, it seems very likely to have been occasioned by the same convulsion of the earth. Now if this earthquake went off southeastward into the *Atlantic*, it would pass considerably to the eastward of St. *Martin*'s; and it is plain that, in fact, it did not reach that island, there being no shock felt there. The motion of the sea, then, was owing to a great agitation raised at a considerable distance, in some part or other of the ocean, where the earthquake passed; and from thence propagated to that island. Nor is the length of time greater than what seems to be necessary for this purpose. The earthquake itself, at the rate it moved with us, would be some hours in going from hence to the latitude of St. *Martin*'s. For it has been already observed, that it's progress was slower than that of sound; and sound would be about two hours and a quarter in moving to such a distance. The rest of the 9^h might well be spent in conveying this motion excited in the water, from the place where it was excited, to St. *Martin*'s; for the waves raised hereby cannot be supposed to have moved with near the velocity of sound.

It is to be regretted, that we have had no good accounts of the time, when the earthquake happened at different places on the continent, whose longitudes and latitudes are known; which might have enabled us to form a more exact judgement of this matter.

A very ingenious Gentleman having given it as his opinion, 'that the earthquake seems to have come from the S. W,' I think my self obliged to consider what he has offered upon this point. His words are, 'Most accounts we have from the S. W. make the shock considerably sooner there, than we had it here, and than it was felt to the N. E. of us; so much sooner, that the difference of longitude will scarce help us at all in accounting for it:' In which, as I humbly conceive, there is a mistake. For, allowing the difference of longitude, the expressions used in the accounts from the S. W. are precisely such as might be expected, in case the earthquake was felt there at the same instant that it was here. I suppose it is sufficiently

Having made the observations propos'd, let us now attempt to

ently evident from what has been said above, that the earthquake began with us at $11'\frac{1}{2}$ after 4, within a very few seconds of time. Now our meridian is to the E. of *New-Haven* about $8'\frac{1}{2}$;—of *New-York*, about $12'\frac{1}{2}$; and of *Philadelphia*, about $17'\frac{1}{2}$. Consequently, when it was $11'\frac{1}{2}$ after 4 with us, it was $3'$ after 4 at *New-Haven*; and it wanted a minute of 4 at *New-York*. Now the accounts from both those places are, that they had the earthquake *about four* o'clock. Again; at this same moment of time, it wanted $6'$ of 4 at *Philadelphia*; and their account is, that the earthquake happened *between three and four* o'clock.

I do not remember any account of the time, from the N. E; unless that from the vessel on the *Atlantic*, 70 leagues E. of *Cape-Anne*, may be reckoned for one. The people on board this vessel are said to have felt the shock at *half past four*. Now 70 leagues on the parallel of *Cape-Anne* make the difference of meridians to be $19'$. So that when it was $11'$ after 4 with us, it was exactly *half past four* at that vessel.

Thus, all these accounts agree perfectly with the supposition, that the earthquake was felt at the same time in these several places. But I am far from affirming, that it really was so. I am indeed inclined to think, that there was no great difference in the time of it's being felt here, and in the places to the S. W. For it seems most likely, that it reached at the same time to all places which lay in a line perpendicular to it's course. So that if it came from about the N. W, as I believe it did, I should think, that all the places beforementioned, *Boston*, *New-Haven*, *New-York*, and *Philadelphia*, which bear almost S. W. and N. E. from each other, must have felt it nearly at the same time. I was careful to note the time, when we had it, as exactly as I could; in hopes that, by comparing it with the like accounts from distant places, we might be able to judge, both of the course of the earthquake, and the velocity of it's progress. But all the accounts of the time, which I have yet seen, are so very lax, that no just conclusions can be drawn from them, with respect to either of these points. To establish any thing of this sort, the minutes at least, if not the seconds, should be made certain.

It is worthy of remark, that two, of the five great earthquakes which we have had, have gone nearly in the same track, as this last did. The first of all, which was on *June* 2. 1638. is said by the Historian, to ' come from the northward, and pass southward.' By the description given of it, it was very much like our late earthquake, only not quite so violent. ' The ' noise and shakes of the earthquake, *October* 29. 1727. seem'd,' it is said, ' to come from the northwestward, and to go off southeasterly; and so the ' houses seemed to reel.' As to the great earthquakes of 1658 and 1663, we have no account of the course which they went in. But from the other three, it may be reasonably conjectured, that the centre, or place in which our earthquakes originate, lies in some part of *Canada*; or perhaps beyond it.

to trace out the causes of these great phænomena.[g]

That the agents, which are able to produce effects so extraordinary as those before recited; which can heave up such enormous masses of matter, and put into the most vehement commotion vast tracts of land and sea, of many hundred miles in extent;—that the agents, I say, which can do all this, and more, must be very powerful, will not admit of a doubt. Now we know of nothing in nature more powerful than the particles of certain bodies converted into vapor by the action of fire. Fire then, and proper materials for it to act upon, it is probable, are the principal agents in this affair. And what greatly strengthens the probability is, an observation before-mentioned, that those countries, which have burning mountains, are most subject to earthquakes; and that these mountains rage with uncommon fury, about the time when the circumjacent countries are torn with convulsions;—an argument this, that the eruptions of such mountains, and earthquakes, are owing to one and the same cause. But we must be more particular.

1. The

---

[g] Several accounts of the late earthquake having been already published by different hands, particularly very full and distinct ones by the Rev. Doctors Chauncey and Mayhew; I think it needless for me to add any thing farther upon this head, except the following short notes.

1. Dr. Mayhew speaks of 'such a great white frost being upon the 'ground in the morning, as he had not observed for many years past.' I took notice of it likewise; and, when it was melted, measured it; and found that it covered the ground to the depth of $\frac{17}{1000}$ parts of an inch; which is almost double of any white frost we have had for seven years past, and about 5 or 6 times as great as we commonly have.

2. The barometer and thermometer underwent no alteration at the time of the earthquake: Only, my barometer, which has an open cistern of quick-silver, was so agitated, that part of the quick-silver was dash'd over the sides of the cistern, and scattered upon the floor. This cistern was a cylindric cup, whose sides were an inch higher than the surface of the quick-silver.

3. I have received undoubted intelligence from a Gentleman in that neighbourhood, that the report we had of 'a prodigious chasm made in the ground at 'Newington in New-Hampshire, of 60 rods in length, and 2 feet in breadth,' is a pure fiction; and that it had no other foundation than this (if this indeed can be called a foundation for such a story) that a stone fence in that town, standing upon a sandy ground, was shaken down by the earthquake; and the force of the falling stones, having beat up the sand on each side, made some appearance of an hollow in the middle.

1. The earth is not solid throughout, but contains within it many large holes, pits and caverns; as is agreed by all Natural Historians. There are very probably also long, crooked, unequal passages, which run winding through a great extent of earth, and form a communication between very distant regions. [h] Some of

---

[h] The observation here made concerning the great length of subterranean canals has received a new and strong confirmation by the account, inserted in note [f], of the agitation of the sea at St. *Martin*'s; and a stronger still, by another account we have more lately received of a surprizing flux and reflux of it at *Barbadoes*, on the first of *November* last, at two o'clock in the afternoon; which was about 6 hours after that terrible earthquake, which made such havock at *Lisbon*, and the neighbouring parts of *Portugal* and *Spain*. This account, as it was published in our News-papers, is as follows. *Extract of a Letter from* Barbados, Nov. 13, 1755. " The first of this Instant, about 2 o'Clock, P. M. the Tide having then ebbed one Hour, in a stark Calm, there was, of a sudden, to the great Surprize of all the Inhabitants, a Flux of Water, that rose to the Height of 5 Feet and upward, and as suddenly it returned to the Sea; this Flux and Reflux was perform'd every 6 or 7 Minutes, and so continu'd 'till 10 o'Clock, tho' the Violence abated gradually from 5 o'Clock: So that in that small Space of Time, the Sea flux'd and reflux'd, upwards of 60 Times." Extraordinary commotions in the sea were likewise observed, the same day, on the western coasts of *Europe* and *Africa*, and at a considerable distance from the shore. One ' vessel in Lat. 25. 30
' N. and Long. 16. 15 W. from *London*, about half an hour after 9 A. M.
' was violently agitated for the space of 3 minutes by an earthquake, tho'
' not damaged,' as we were told in our News-papers. This vessel was then under a meridian half an hour to the W. of *Lisbon*, and at the distance of about 1000 miles from it; and the time, when it felt the shock, was an hour earlier than they had it at *Lisbon* and *Cadiz*; and therefore, if this was the same shock as that, it was an hour in moving from the place of this vessel to those cities. ' At *Lisbon*,' we are told, ' half an hour after the shock,
' which came between 10 and 11 o'clock in the forenoon, and continued
' about 2 minutes, the tide came in and arose about 20 feet higher than
' common, by which a few people were drowned. And at *Cadiz*, that soon
' after the shock, which was at 11 in the forenoon, and lasted about 3
' minutes, they saw from the shore a heavy sea (about half a mile distance)
' coming on, which did considerable damage to the shipping, and when it
' broke, made great destruction in the city :—That about a quarter of an
' hour after, there came a second sea as awful; and about the same space
' after, came a third more awful, which did prodigious damage; and great
' numbers of people were lost.' These extraordinary agitations in the ocean, which all happened on the first of *November*; and another of the
same

of these cavities are dry, and contain nothing but air, or the fumes of fermenting minerals; in others, there are currents of water.

2. This globe is a very heterogeneous body. Besides the two grand divisions of it into solid and fluid parts, each of these is again divisible into an indefinite number of others. Although our knowledge of the earth reaches but a little way below it's surface, yet so far as we have penetrated, it appears to be a *compages* of a vast variety of solid substances, ranged in a manner which to us seems to have not much of regularity in it. Here we find earths, stones, salts, sulphurs, minerals, metals, &c. and a great number of inferior species under each of these general heads, blended and intermingled with each other. Many of these are combustible, or of a texture proper to be turned by fire into flame and vapor. And besides the pure elementary water, if there be any such, the aqueous parts of the globe receive

peculiar

same kind on the 18th, which was observed at St. *Martin*'s, as is related in note f ; we shall endeavour to account for in note aa. p. 24.

Here now we have an earthquake extending it's effects across the *Atlantic*. An amazing distance truly for an earthquake ! And such as I do not remember to have met with any account of, in history. Our earthquake reached 1900 miles ; whereas this reached almost twice as far.

But though this latter earthquake was of greater extent than the former, perhaps it did not much, if at all, exceed it in the degree of violence. Indeed, the damage done by it at *Lisbon* appears to have been vastly greater than any sustained by us ; though we have not yet had an exact account of the particulars. But this was probably owing, not altogether to the greater violence of the shock, but partly, if not chiefly, to their manner of building. I am informed by a Gentleman, who is well acquainted with that city, that the buildings are very high, many of them being of 7 stories ; and chiefly of stone, laid in clay. A poor cement indeed ! which can never bind stones together with any tolerable firmness. And on supposition that their earthquake was of the same species with ours, that is, *undulatory*, the great height of those buildings exposed them to much greater agitations, than if they had been lower. (See p. 10.) It is no wonder then, if their houses were shaken to pieces, though the shock were not more violent than what has been felt in other places, where much less damage has been done. An earthquake that could give so smart a blow, as ours did to some windvanes (See note d) must be sufficient, one would think, to demolish such a city in a few minutes ; so as scarcely to leave one stone upon another. Had *Boston* been built in that manner, probably at this day it would have been little better than an heap of ruins.

peculiar tinctures from the beds and veins through which they run; so that perhaps there may be almost as many sorts of waters, as there are of solid substances. Thus, some waters are charged with sulphureous particles; some, with particles of iron; and others, with those of other minerals. And the subterraneous rivers and streams, thus impregnated with different particles, may, by their confluence, produce an almost infinite variety of mixtures in the earth. Probably, this promiscuous disposition of materials, in the bowels of the earth, may be necessary to the growth of bodies in it; for in the judgement of some of the most eminent philosophers, particularly the excellent Mr. BOYLE, even the hardest, inorganized bodies, as stones, metals, &c. do, in their proper way, grow within the earth, as truly as vegetables grow on it's surface, or animals in their parent animals. And to the same end seems to conduce

3. The heat within the bowels of the earth. Heat, it is well known, is a grand agent in most natural productions; and the inner parts of the earth are sufficiently furnished with it. Some parts indeed, as the vulcanos, are actually on fire and burn; but there is, moreover, an heat without flame, diffused through the interior regions of the earth. This is evident from the instance of hot springs, and from the warmth which is always found at great depths, as in the bottoms of mines.

4. There seems to be an inexhaustible source of this heat in the attractive powers, which Sir ISAAC NEWTON has shewn to belong to the particles of matter. For, heat consisting in a peculiar kind of intestine motion of the parts of bodies; whatever tends to produce this motion in bodies, will cause them to grow hot. Now such a motion may be produced, by the particles of different bodies rushing together, in virtue of their attractive powers; of which that great man has given a very copious collection of instances in the 31st *Question* at the end of his *Opticks*, whither I must refer you. In some of them, not only a very sudden and violent heat, but an actual flame, is produced, by the bare mixing of two cold bodies together; and that, even without the presence of the air, which we find absolutely necessary to our culinary fires. So in a remarkable experiment, (first made, I think, by Dr. SLARE) when a certain ' compound
' spirit

' spirit of nitre is poured on half it's weight of any ponderous
' oil of vegetable or animal substances, the liquors grow so
' very hot in mixing, as presently to send up a burning flame.' ⁱ
At the first trial, so small a quantity, as ' a drachm of this spirit,
' being poured upon half a drachm of such oil, *in vacuo*, to see
' what effect would ensue; the mixture in the twinkling of an
' eye made a flash like gun-powder, and blew up the exhausted
' receiver, whose diameter was six inches, and depth above
' eight; all who were present being astonished at the unexpected
' event.' There is also so strong an attraction between iron
and sulphur, that ' even the gross body of sulphur powder'd,' continues Sir ISAAC, ' and with an equal weight of iron filings and
' a little water made into paste, in a few hours grows too hot to
' be touch'd, and emits a flame. And by these experiments
' compared with the great quantity of sulphur with which the
' earth abounds, and the warmth of the interior parts of the
' earth, and hot springs, and burning mountains, and with damps,
' mineral coruscations, &c. we may learn that sulphureous
' steams abound in the bowels of the earth, and ferment with
' minerals, and sometimes take fire with a sudden coruscation
' and explosion.' But to set this curious doctrine in it's full
light, it would be necessary to repeat that whole *Question*;
which indeed highly deserves it, would the time permit. For it
would then appear, that there is a very great variety of bodies,
which being mixed together, produce so strong an effervescence,
as to emit inflammable fumes. Thus, to mention one instance
more, when iron is dissolving in a mixture of oil of vitriol and
common water, there instantly arises a great heat and violent
ebullition, with fumes copiously exhaling; which are so very
inflammable, that, being set on fire, they go off at once like a
gun, with a great explosion. Having thus seen what a perpetual
source of heat there is in these powerful, active principles, continually

---

ⁱ *Newton*'s Opt. p. 353. This compound spirit of nitre was made by distilling equal parts of nitre and oil of vitriol. The oil used in Dr. *Slare*'s first experiment was that of carraway seeds. A full account of these experiments, by Dr. *Slare* himself, may be seen in *Lowthorp*'s Abridgm. Vol. III. p. 352—366.

tinually operating within the bowels of the earth; let us next inquire, what effects may be expected from it. Therefore

5. It is a known property of heat to expand bodies, to rarefy them, and enlarge their dimensions; and, when raised to an higher degree, to separate their parts, and make them fly from each other; as in some measure appears already from the instances mentioned under the foregoing article. This effect heat has upon solid as well as fluid bodies;—upon the hardest, as well as the softest. It is observable here, that such particles as cohere by the strongest attraction, do most forcibly repel one another, when they are once separated by heat. And when the heat is intense, and the particles of the heated body are prevented from flying away, till they become thoroughly hot; it will require very strong vessels to hinder their bursting forth with a violent explosion. Thus, a single drop of common water, inclosed in a glass bubble, and laid upon the fire; as soon as it becomes hot, will burst the bubble, with a report scarce inferior to that of a pistol. And water in larger quantities has been heated to that degree, as to rend in sunder very strong vessels of iron, in which it has been endeavoured to be confined. What the consequence then would be, of a great body of water's suddenly making it's way into a flaming cavern, whose sulphureous or bituminous fires are not extinguished but inraged by water; and of it's being there, almost instantaneously, converted into vapor; your own imaginations may easily represent to you. This, it is very likely, has sometimes been the case with respect to those famous vulcanos, *Ætna* and *Vesuvio*; both which border on the sea. You see here what water may do; but there are many other bodies, which cohere more strongly, as sulphur and nitre, for example, whose vapor is still more powerful than that of water. This is evident from the composition of gun-powder; a very small quantity of which, when turned into vapor, every one knows, is able to remove any obstacle that opposes it's expansion, and to burst the firmest rocks. The paste abovementioned, made of powdered sulphur and iron filings, if put a few feet under ground, will by degrees cause the earth over it to heave and crack, to let out the flame; thus making an artificial earthquake. And therefore, if a water,

saturated

faturated with fulphureous particles, fhould in it's paffage under ground foak into a large bed of iron ore, or a ftrong chalybeate water into a bed of fulphur; the mixture would doubtlefs perform *in great*, what this experiment does in miniature. A vitriolic water mixing with iron, if in fufficient quantities, would be followed with the like effect. But no mixture of this nature appears fo furprizing as Dr. SLARE's, which did not require fo much as the prefence of the air to inkindle it. From this fmall mixture, which was but one drachm and an half, a force was generated in an inftant, by the mutual collifion of thofe active liquors, far exceeding the weight of the atmofphere that preffed down the receiver; which in that experiment amounted to about 420 pounds. I fay, *far exceeding*; for the receiver was not barely raifed, but ' with a much greater force ' blown up,' as Dr. SLARE expreffes it. We have no right indeed, that I know of, to fuppofe all the fame forts of bodies, both folid and fluid, in the bowels of the earth, as chemiftry has furnifhed us with; but feveral things induce us to believe, that in thofe dark recefles, impenetrable as they are to mortal eyes, bodies are prepared, by a kind of natural chemiftry, which very much refemble many of our chemical preparations, and are poffeffed of the fame effential properties. To be fure, there are not wanting, in thofe lower regions, all the degrees of heat, which a Chemift could defire for any of his proceffes. If then two bodies, of the like nature as the fpirit and oil ufed in this experiment, fhould be mingled together in due quantity, though in the clofeft fubterraneous vault, which neither contained any genuine air, nor could admit any; I need not fay, that an earthquake muft be the confequence.

You have now, I fuppofe, before you the general caufes of earthquakes. You have feen that there are in the bowels of the earth inflammable materials, of various kinds, and in large quantities; fome in the form of folid or liquid bodies, and others in that of exhalations and vapors; that there are alfo powerful principles conftantly at work, which are capable of inkindling thefe materials into an actual flame; and that the vapor generated from fuch flame will endeavour to expand it felf on all fides

with

with immense force. If now these inflammable vapors be pent up in close caverns, so as to find no vent till they are collected in a large quantity; so soon as they take fire in any part, the flame will spread itself, wherever it meets with materials to convey it, with as great rapidity, perhaps, as it does in a train of gunpowder; and the vapors produced from hence will rush along through the subterraneous grottos, as they are able to find or force for themselves a passage; and by heaving up the earth that lies over them, will make that kind of progressive swell or *undulation*, in which we have supposed earthquakes commonly to consist; and will at length burst the caverns with a great shaking of the earth, as in springing a mine; and so discharge themselves into the open air. [22] These vapors may possibly sometimes infect

[22] The extraordinary commotions of the sea, related in the foregoing notes, having happened within a few hours of the great earthquakes; one of which shook *Spain* and *Portugal*, and the other, *New-England* with some of the neighbouring parts of *America*; will naturally be ascribed by every body to those earthquakes, or at least to the same causes as those earthquakes are. Now for my part, I can hardly persuade my self, that the bare agitations of the earth, at those times, could be great enough, to put the sea into such vehement commotions, as it appears to have been in, by those relations. To account for these things satisfactorily, it seems to me that we must have recourse to such an *eruption* of the vapors which caused those earthquakes, as is spoken of in the passage above. At those times, these furious vapors, impatient of restraint, must have continued to drive along through their subterraneous passages, till they found some place, where the top of the caverns, which contained them, was not of sufficient strength to confine them; and there they would burst out of their dungeons, and spring up into day. The eruptions, which caused those uncommon motions of the sea that surprised the inhabitants of *Barbadoes* and St. *Martin*'s, were very probably made in the *Atlantic* ocean, to the eastward of the *West-India* islands, and near the same latitudes; and, until we hear whether the like commotions were observed elsewhere, about the same time, we may well enough suppose, that the places of these eruptions were nearer to those islands, than to any other land. Now to assist the Reader in forming a conception of what probably passed in the ocean on those great occasions, I shall lay before him two or three facts, relating to the force of fired gun-powder, and the resistence of water; taken from the before cited work of Dr. DESAGULIERS, pp. 420, 421. The Dr. tells us, that ' being with several other persons in a great barge upon the ' *Thames*, playing off some fire-works on a rejoicing day, it happen'd that a
' water-

fect the air, and bring on pestilential distempers, which have been said to be consequent upon great earthquakes. Not that I can

'water-rocket (whose property is to go under water several times and rise
'again, and at last burst on the top of the water) came up, when it was
'ready to burst, under the stern of the barge, being thereby prevented from
'coming up to the surface of the water; and bursting where it was, gave
'the barge a great shock, and sensible lift, though there was much less than
'an ounce of powder to make it. Another of these rockets in it's last rise,
'stopping under the middle of a smaller barge, broke there, and made so
'great an hole in the barge's bottom, that in a very short time the barge was
'half full of water. After this, to try the effect of the explosion of gun-
'powder under water; the Dr. loaded a water rocket, so that it should
'break under the water, and having set fire to it, threw it into a pond, that
'covered an acre of ground: And so great was the shock, that several per-
'sons, who stood round the pond, felt it like a momentaneous earthquake.'
Now if so inconsiderable a quantity of powder, fired just under the surface of water, could have such effects as these; what must have been the commotion, when the vapors, which were able to shake such great extents of land and sea, as we are sure were shaken in these earthquakes, made their way, with united force, through the vast body of water that lay over them? No doubt, the water foam'd, and boil'd, and raged with inconceivable fury, and was agitated into overgrown, mountainous waves. The first effect of the eruption probably was, that all the water, which lay directly over the spot where the bottom of the ocean gaped to let out the vapors, was blown right up, almost like a compact body, to a great height in the air. The bottom doubtless closed again as soon as the vapors were discharged; but there must have been a pit or cavity left in the ocean, in the place deserted by the water:—of what dimensions it is impossible for us to say; though, from what followed it seems, they must have been very considerable. The next step would be, that the neighbouring water would rush in from all sides to fill up the vacuity; first, from the nearer parts; and then by degrees from the remoter; and by that means form a spacious concave all around, on the surface of the ocean; the centure of which would be this pit. The motion of the water descending to fill such a pit, was what, I suppose, might draw off the water from the shore of St. *Martin*'s; which was the first circumstance observed there. The water, by thus descending to fill the pit, having fallen below it's proper level, would next be raised above it, erecting itself into a mountain over the place where the pit was made; and then by falling and rising alternately in this place, would communicate an undulatory motion all around it: And the waves thus excited would be more numerous, and of greater breadth, as the dimensions of the pit first made were larger. Mean time, the water, thrown up at the beginning in

I can give credit to all the reports of this sort, which have been handed about; many of them having been propagated by writers of an astrological turn, who have been as ready to attribute distempers to the configurations of the planets, and to the appearance of comets, as to earthquakes.

By this time, enough has been said, I should think, to convince you, that the earth contains within itself the seeds of earthquakes in great abundance. And all these things being considered, it may seem rather a wonder that we have no more earthquakes, than that we have so many. The causes of earthquakes are incessantly at work; and although it may require a course of years [bb] to ripen the proper materials to that pitch, as that

a body into the air, would by it's weight fall down in cataracts, and add greatly to the confusion. A motion like this, once begun, must needs be propagated to very considerable distances, before it could be intirely lost; and that to a degree sufficient, I should think, to cause such great waves, and to such a number, as were observed at the places before-mentioned. Whether this, or something like this, might not probably have been the process of these extraordinary scenes in the ocean, I submit to the judgement of the Reader. And if he shall be of this opinion, he will doubtless make a pause, and reflect on the great goodness of HEAVEN, in causing the vapors to break forth in the ocean;—a place, where they could do the least hurt. The effects which must have followed, had these impetuous blasts been directed against the foundations of a great and populous city, his own imagination will paint to him in livelier colors than I can pretend to do.

It is necessary to suppose, at least two such eruptions in the *Atlantic* on the first of *November*. For it is utterly incredible, that the waves excited by one only, however great in degree, or in whatever part of the ocean; it be supposed, could reach to places at such distances from each other, as *Cadiz* and *Lisbon* are from *Barbadoes*. The inundations at *Cadiz* and *Lisbon* were probably occasioned by one and the same eruption, which, by the violence of them, seems to have been at no great distance from those cities. But that at *Barbadoes* must have been occasioned by another eruption, happening a few hours later; and seems, by the expressions used in the relations, to have fallen far short of those in *Europe*, in point of violence. From whence we should argue, that this eruption happened at a considerable distance from that island: Because, from the great number of waves, or fluxes and refluxes, we must argue, that the commotion in the centre or place of eruption, which excited those waves, was vastly great.

[bb] Besides several lesser earthquakes, there have been five great ones in *New-England*, since the arrival of the *English* in 1620; which, at a medium, is one

that they can force for themselves a paſſage thro' the earth; yet it is reaſonable to expect, that they will from time to time be collected in ſuch quantities, and ferment to ſuch degrees, as to make theſe exploſions unavoidable. As therefore our globe has been ſubject to ſuch concuſſions from the earlieſt accounts of antiquity, we have no room to doubt but that it will continue to be ſo, as long as the preſent frame of nature ſubſiſts. For this we may be aſſured of, that though that impriſoned vapor be diſcharged into the open air, which, by its ſtruggles to eſcape, has cauſed an earthquake; yet the fermenting minerals, from which it was generated, will be continually ſupplying new quantities of the ſame; and even thoſe very minerals may from time to time be re-produced, as they are conſumed; as was before obſerved. Thus we ſee, that in the very ſtructure and conſtitution of this globe, proviſion has been made to continue theſe agitations of it, at proper intervals of time, during the whole period of it's exiſtence in it's preſent form; and that in every climate, from the equator to the pole. This ſuggeſts a reflection, with which I ſhall cloſe the preſent diſcourſe. It is this: That

Though theſe exploſions, and conſequent concuſſions of the earth, have indeed occaſioned moſt terrible deſolations, and in this light may juſtly be regarded as the tokens of an incenſed DEITY; yet it can by no means be concluded from hence, that they are not of real and ſtanding advantage to the globe *in general*. Multitudes, it is true, have at different times ſuffered by them; multitudes have been deſtroyed by them; but much greater multitudes may have been every day benefited by them. The all-wiſe CREATOR could not but foreſee all the effects of all the powers he implanted in matter; and, as we find in innumerable inſtances (and the more we know of his works, the more ſuch inſtances we diſcover) that he has eſtabliſhed ſuch laws for the government of the world, as tend to promote the good of *the whole*,

---

one in about 27 years. But they have not happened at equal diſtances of time. About the year 1660, they were pretty frequent; but I find no mention of any from 1670 till that memorable one in 1727; between which there was an interval of 57 years.

whole, we may reasonably presume, that he has done it in this case as well as others. To me, at least, the argument on this side the question, drawn from the general analogy of nature, appears to have more force, than any that I have seen offered on the other. For there is nothing, however useful, however necessary, but what is capable of producing, and in fact has produced, damage, in single instances. It were endless to particularize here; I shall therefore only mention one or two things by way of specimen. The power of gravity,—a power of such indispensable importance, that without it the system of nature could not subsist a moment, has yet proved the destruction of multitudes. The wind, so necessary for the purposes of navigation, as well as to purge the air, which would otherwise stagnate and putrefy,—how often has it risen to such a pitch, as to overthrow houses, and wreck vessels? by which means thousands have perished. Even thunder and lightning, which, next to earthquakes, are the most terrible phænomena of nature, are yet universally allowed to be necessary to free the atmosphere from a certain unwholsome sultriness, which often infects it. Other instances of the like sort I leave to your own reflections: and would rather observe, that the world is governed by *general* laws; and general laws must, from the nature of them, be liable sometimes to do hurt. However, laws of this sort are sufficiently vindicated, not only as *wise*, but as *good*, if upon the whole they produce a *maximum* of good; (to borrow an expression from the Mathematicians;) and this, it is in the highest degree probable, all the laws of nature do. It may be added, that as in the animal body, the evacuations, which are of absolute necessity to maintain life and health, do yet sometimes run to such extremes as to prove mortal; so in like manner, these explosions of subterraneous vapor, whose effects have sometimes been so fatal, may, notwithstanding this, be highly conducive, and even indispensably necessary, to the good of this globe in general. The explosions themselves, as well as the laws, in consequence of which they are produced, may be necessary on various accounts;

and

and particularly to the carrying on the more secret and noble works of nature within the entrails of the earth. " Let me dilate a little on this matter.

By the incessant action of gravity and other attractive powers, and by the perpetual consumption of fluids, the earth becomes continually more and more hard, compact and dense. Now an openness or looseness of contexture, to a certain degree, in the earth, is necessary to carry on the operations of nature within it. So that on the supposition that mineral, metalline, and other subterraneous bodies grow within the earth, it should seem that the earth must become gradually less and less fit for the production of them. Since then the direct, immediate, and most general effect of earthquakes is, by shaking to loosen and disunite the parts of the earth, and to open it's pores, it seems agreeable to reason to infer, that this is the end *primarily* aimed at in these concussions. But you will take notice, that I speak here only of *physical* or *natural* ends. For though I make no doubt, that the laws of nature were established, and that the operations of nature are conducted, with a view, *ultimately*, to *moral* purposes; and that there is the most perfect coincidence, at all times, between GOD's government of the *natural* and of the *moral* world; yet it would be improper for me to enter into these disquisitions at this time, since my province limits me to consider this subject, only in the relation which it bears to *natural philosophy*. It is in the *physical* sense alone that I say, the disjoining the parts of the earth, and opening it's pores, may be the end primarily aimed at in earthquakes, as such mutations in the earth may from time to time become necessary to the production of subterraneous bodies;

" It is not impossible that some may think it strange to have any thing said, that seems at all to abate the horror which many people have of earthquakes, as if *all* of them were *nothing but scourges* in the hand of the ALMIGHTY; and may be fearful, lest the cause of religion should be disserved hereby. But of this there is not, in my apprehension, the remotest danger. The idea here exhibited, while it exalts the wisdom and goodness, does not in the least detract from the majesty, or from the justice, of GOD. And the terror, which an earthquake never fails to carry with it, will be sufficient to secure the interests of religion, so far as they are to be secured by the influence of fear; even though such a phænomenon be represented in the most favorable light that truth will admit of.

bodies; and perhaps this end could not be effectually anfwered by lefs forcible methods. This point may receive fome light, if not proof, from the operations of agriculture. We find it neceffary, by ploughing, digging, &c. to break the clods of the ground, to comminute and even pulverize it, in order to fit it for the purpofes of vegetation; and we find it neceffary to renew thefe labors every year. Now the ufe and tendency of thefe artificial operations may bear fome analogy to thofe of the greater operations of nature, which we are fpeaking of. And indeed, it is not in the leaft degree improbable, that fuch a loofening of the parts of the earth may promote even the growth of vegetables on it's furface, as well as of minerals in it's bowels; it being now well known, that all vegetables, the fmaller as well as the larger, fhoot fome fibres of their roots to vaftly greater depths, than thofe to which any of our inftruments of tillage ever penetrate. This, it is likely, may be one reafon of the wonderful fertility, for which *Ætna* and *Vefuvio* have been fo generally and fo highly celebrated. Again; it may be neceffary now and then, to have fuch fubterraneous vapors, as are generated by fermentation, difcharged up into the air; as their continuance below, in the caverns of the earth, might be an impediment to thofe important proceffes which are there carrying on. But thofe very vapors, which might obftruct fome forts of natural proceffes, while below the furface of the earth, may as much advance others, when above it. We know that in many cafes of the fermentation of bodies, efpecially of fuch denfe ones as falts and minerals, air is plentifully *abforbed*; and that in many others, it is as plentifully *generated*: So that great part of the exhalations thrown out by earthquakes may be true, permanent air, and defigned to recruit what has been abforbed by bodies here on the furface. And perhaps the grounds, on which the great NEWTON founded his ' fufpicion, [dd] that the fineft, the moft fubtile,
' and moft fpirituous parts of our air, and thofe which are moft
' neceffary to maintain the life of all things, come chiefly from
' the comets'; may equally fupport another fufpicion, that fome
such

[dd] *Newton's* Princip. p. 515.

such particles of air may be derived also from subterraneous eruptions. For among the almost infinite variety of particles which are thrown out of the earth in these eruptions, it is most likely, that if some are noxious, others will be salutary. It may also be necessary from time to time to have the subterraneous streams diverted from their former courses into new ones: partly, that different places in the lower regions may be watered by them; and partly, that the waters themselves, by passing through different beds or chanels, may alter their properties, ᵉᵉ and convey new tinctures to different places.

But however these things may be; whether all the foregoing conjectures be well founded, or not: If these explosions and concussions be, as it is next to certain that they are, the necessary and inevitable consequences of such laws of nature, and such powers in matter, as our globe could not well subsist without; this ought to silence all the complaints of those who suffer either loss or terror by them; as well as all the objections, which men of sceptical minds have been disposed to make, upon this head, to the order of Providence. It ought, in reason, to do this, though we should never be able to point out all the particular advantages resulting from them. For, it is plain, they may be beneficial in a thousand other ways, than we, short-sighted mortals, may pretend to guess at.

To sum up all in a word. This is a MIX'D state; in which there is such a variety of purposes, *natural* as well as *moral*, in prosecution at the same time, that there may be nothing, perhaps, in the material world, that is simply and absolutely *evil*;— nothing, but what, under the direction of infinite wisdom, power and beneficence, is, in some or other of it's consequences, productive of an over-balance of *good*.

Upon the whole. How 'wonderful in counsel', how 'excellent in working' is that BEING, who can bring good out of the greatest evils; and can answer intentions, the most widely differing, by one and the same dispensation of His providence!

ᵉᵉ Some wells near me have had the quality of their waters much mended since the earthquake.

The END.

APPENDIX,

[*Editor's Note:* The appendix, "Concerning the Operation of Electrical Substances in Earthquakes; and the Effects of Iron Points" (pp. 32–38), has been omitted here.]

Part II
FOSSILS

Editor's Comments on Papers 3 Through 7

3 **ANONYMOUS**
   Description of a Remarkable Tooth, in the Possession of Mr. Peale

4 **JEFFERSON**
   A Memoir on the Discovery of Certain Bones of a Quadruped of the Clawed Kind in the Western Parts of Virginia

5 **BARTON**
   Facts, Observations, and Conjectures, Relative to the Elephantine Bones (of Different Species), That Are Found in Various Parts of North America

6 **CLEAVELAND**
   Account of Fossil Shells, with the Author's Reasons for Attending to the Same

7 **LE SUEUR**
   Observations on a New Genus of Fossil Shells

Paleontology in America began with isolated discoveries of fossil bones and shells in the eighteenth century. Giant bones of extinct quadrupeds were the first fossils to attract much attention and the earliest printed account of an American vertebrate fossil may be the 1714 report by Cotton Mather on bones found several years earlier in New York. This paper, which appeared in the *Philosophical Transactions of the Royal Society*, included descriptions of teeth and a leg bone presumed to belong to an antediluvian giant man, as mentioned in the Scriptures.

Additional discoveries of fossil bones were made by many investigators and publicized in brief notes in early American periodicals. David Rittenhouse (1732-1796), a noted American astronomer, instrument maker, and second president of the American Philosophical Society, discovered a "remarkable tooth in the possession of Mr. Peale"* that

---

*Charles Willson Peale's excavations of fossil bones gained him international renown, and his complete mounted fossil skeleton of a mastodon was the first such display

differed from "other grinders which have been brought from the Western Country" (Paper 3). The accompanying plate clearly illustrates the grinder of a mammoth. Almost all the eighteenth-century elephantine fossils reported from the Colonies were of the mastodon (Greene 1959), and Rittenhouse's find may have been the first and only pre-1800 description of a mammoth in North America. The prevalence of mastodon teeth in America puzzled European workers who had found mammoth remains on their continent. The common usage by American workers of the term "mammoth" for any large fossil elephant further confused the record of vertebrate remains in North America. The Rittenhouse discovery, which would have demonstrated the existence of two distinct elephant-like animals in the Americas, was apparently little noted by workers either domestic or foreign.

An important discussion of fossil bones, written by another president of the American Philosophical Society, was Thomas Jefferson's (1743–1826) "Discovery of certain bones of a quadruped of the clawed kind in the western parts of Virginia" (Paper 4). This contribution was especially important, for Jefferson's prestige as statesman, author, and president of the United States added a glamour and legitimacy to the study of these ancient relics. The bones, to which Jefferson assigned the name *Megalonyx* or "great-claw," actually belonged to an extinct giant sloth, rather than a relative of the lion, as he suggested.

Jefferson's discovery provided data relevant to at least two debates on the nature of animals in North America. First, the discovery of this giant animal strengthened the evidence refuting Marquis de Buffon's theory that American animals were generally smaller in stature than corresponding European varieties. Jefferson documented that the American moose, deer, mammoth, and *Megalonyx* were all larger than comparable European species. Jefferson's conclusions were a tribute to his logic and his pen:

> Are we then from all this to draw a conclusion, the reverse of that of M. de Buffon. That nature, has formed the larger animals of America, like its lakes, its rivers, and mountains, on a greater and prouder scale than in the other hemisphere? Not at all, we are to conclude that she has formed some things large and some things small, on both sides of the earth for reasons which she has not enabled us to penetrate; and that we ought not to shut our eyes upon one half of her facts, and build systems on the other half.

A second and perhaps more significant issue considered by Jefferson was that of animal extinction: "What is become of the great-claw?" It was not evident in 1799 whether these giant animals were extinct or merely living in remote country. Jefferson, an avowed empiricist, pre-

---

outside Europe. Though the note (Paper 3) is unsigned, Peale may well have been the author.

ferred to believe that bones represented living animals until proven otherwise. Tales of unusual animals were not uncommon in reports from the western country, and some of these accounts might be interpreted as evidence for a lion-like animal in the West. Jefferson concluded that "if this animal has once existed, it is probable . . . that he still exists."

By 1806, when Benjamin Smith Barton (1766-1815) presented his "Facts, observations, and conjectures, relative to the elephantine bones . . . of North America" (Paper 5), mastodon or mammoth remains had been found in a dozen states. The numerous isolated finds were gradually providing a systematic picture of the nature and distribution of these extinct mammals. Especially exciting was the discovery of a mammoth with soft parts preserved in a salt lick in Wythe, Virginia. Barton, a physician with strong interests in natural history, was founder and editor of the *Philadelphia Medical and Physical Journal,* which proved a logical forum for his analysis of these fossils.

Barton's article was actually a letter to Georges Cuvier, eminent French naturalist and founder of comparative anatomy. It was Cuvier, more than any other researcher, who convinced the world of the reality of animal extinction. The ever-increasing number of elephantine fossils in North America, with no sign of living species, led Barton to support Cuvier's position. (See, however, "note 2" on page 35.) An important conclusion of Barton's contribution was that the mammoth was herbivorous and not carnivorous as claimed by some previous writers (Peale 1802).

Cuvier's influence on vertebrate paleontology in America was enhanced by publication of an English translation of his *Essay on a Theory of the Earth,* edited and enlarged with observations on American fossil remains by Samuel Latham Mitchill in 1818.* Cuvier's *Essay,* in applying comparative anatomy to the classification of vertebrate fossil remains, emphasized the relationships between morphology and function in an animal's bones. This work, coupled with the numerous fossil bone discoveries of the previous century, placed vertebrate paleontology on a firm footing in America.

Fossil shells were less spectacular than giant bones and teeth, and consequently inspired less comment. Their importance in delineating a region's geological structure, however, was not missed by Parker Cleaveland (1780-1858). Cleaveland, best remembered for his *Elementary Treatise on Mineralogy and Geology* (see Paper 12), was professor of mathematics, chemistry, natural history, and mineralogy at Bowdoin College in Brunswick, Maine. His "Account of fossil shells, with the

---

*This book has been reprinted by Arno Press and is not included in the present volume.

reasons for attending to the same" (Paper 6) is one of the earliest statements by an American of the possible use of fossils in correlating strata in different regions. Cleaveland's descriptions of the shells, however, leave the reader in some doubt as to the exact nature of the animals in question. Few Americans were trained in the identification and description of invertebrate fossil remains prior to 1820.

Charles Alexander Le Sueur's (1778-1846) "Observations on a new genus of fossil shells" (Paper 7) was the earliest precise description of a North American fossil shell, which was appropriately named after the United States' most eminent geologist, William Maclure. Le Sueur's description and beautiful illustration are clearly recognizable as the Ordovician gastropod, *Maclurites*. In spite of these efforts, invertebrate paleontology received no systematic attention in North America until the 1830s, when the articles and monographs of Isaac Lea, Timothy Conrad, Jacob Green, James Hall, and C. S. Rafinesque were published. The needs and opportunities in American paleontology, however, were clearly defined by 1820.

REFERENCES

Greene, J. C. 1959. *The Death of Adam. Evolution and Its Impact on Western Thought*. Ames, Iowa: Iowa State Univ. Press.

Peale, R. 1802. *Account of the Skeleton of the Mammoth, a Nondescript Carnivorous Animal of Immense Size Found in America*. London: E. Lawrence.

BIBLIOGRAPHY

Annan, R. 1793. Account of a Skeleton of a Large Animal. *Am. Acad. Arts Sci. Mem.* **2**:160-164.

Atwater, C. 1820. On Some Ancient Human Bones, etc. With a Notice of the Bones of the Mastodon or Mammoth, and of Various Shells Found in Ohio and the West. *Am. Jour. Sci.* **2**:242-246.

Baker, H. 1746. Account of an Extraordinary Large Fossil Tooth and the Bones of an Elephant, with Reflections Occasioned Thereby. *Am. Mag. and Hist. Chron.* **3**:541.

Barton, B. S. 1807. Additional Facts and Observations, Relative to the Extinct Species of American Elephants. *Phila. Med. Phys. Jour.* **second suppl.**:309-311.

Brongniart, A. 1818. Notice of M. Brongniart on Organized Remains. *Am. Jour. Sci.* **1**:71-74.

Collinson, Peter. 1789. An Account of Some Very Large Fossil Teeth Found in North America. *Am. Mus. or Universal Mag.* **5**:155-157.

Cuvier, G. 1799. Notice Concerning the Skeleton of a Very Large Species of Quadruped, Found in Paraguay, and Deposited in the Cabinet of Natural History at Madrid. *Universal Mag.* **1**:337-339.

Cuvier, G. 1818. *Essay on the Theory of the Earth.* New York: Kirk and Mercein, 431p.

Dunbar, W. 1804. Extracts of a Letter from William Dunbar Esq. of the Natchez to Thomas Jefferson, President of the Society. *Am. Philos. Soc. Trans.* **6**:40–42.

Edwards, T. 1789. Description of a Horn or Bone Lately Found in the River Chemung or Tyoga, a Western Branch of the Susquehanna, About Twelve Miles from Tyoga Point. *Am. Mus. or Monthly Mag.* **4**:42. (*also reprinted in Am. Acad. Arts Sci. Mem.* **2**(1793):164–165.)

Hunter, W. 1789. Observations on the Bones, Commonly Supposed to be Elephant's Bones, which Have Been Found Near the River Ohio, in America. *Am. Mus. or Universal Mag.* **5**:152–155.

Jefferson, T. 1788. *Notes on the State of Virginia.* Philadelphia: Prichard and Hall, ii, 244p.

Meriwether, D. 1803. Particulars of a Remarkable Body of Sea-shells Now Existing in the Interior Part of the State of Georgia. *Med. Repos.* **6**:329.

Peale, R. 1804. A Short Account of the Mammoth. *Lit. Mag. and Am. Regist.* **1**:292–297.

Say, T. 1819–1820. Observations on Some Species of Zoophytes, Shells, etc. Principally Fossil. *Am. Jour. Sci.* **1**:381–387; **2**:34–35.

Turner, G. 1799. Memoir on the Extraneous Fossils, Denominated Mammoth Bones: Principally Designed to Show, that They Are the Remains of More Than One Species of Non-descript Animal. *Am. Philos. Soc. Trans.* **4**:510–518.

West, S. 1793. A Letter Concerning Gay Head. *Am. Acad. Arts Sci. Mem.* **2**:147–150.

Wistar, C. 1799. A Description of the Bones Deposited, By the President, in the Museum of the Society, and Represented in the Annexed Plates. *Am. Philos. Soc. Trans.* **4**:526–531, 2 plates.

Wistar, C. 1818. An Account of Two Heads Found in the Morass, Called the Big Bone Lick, and Presented to the Society, by Mr. Jefferson. *Am. Philos. Soc. Trans.* n.s. **1**:375–380, 2 plates.

Description of a remarkable Tooth, in the possession of Mr. Peale.

THE grinder described in the annexed plate, fig. 2, was found on the banks of Susquehanna, near Tioga, in March, 1786. Another part of the same tooth lay near it, which not being taken away at the time, could not afterwards be found. It differs however, on the masticating surface, from other grinders which have been brought from the western country (of some of which I have also made drawings.) The others had several conical nobs of about one inch and an inch and an half prominency, but in this we find some waving, but little elevated ridges, which part, as well as the nobs of the other teeth, are hard enamel.

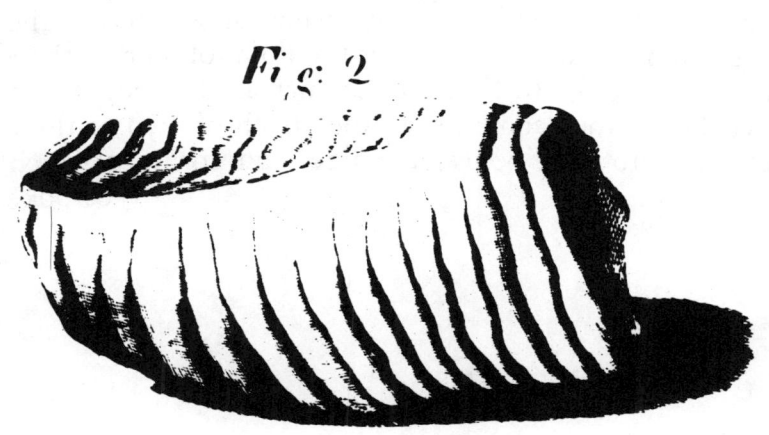

Fig: 2

A MEMOIR ON THE DISCOVERY OF CERTAIN BONES OF A QUADRUPED OF THE CLAWED KIND IN THE WESTERN PARTS OF VIRGINIA

Thomas Jefferson

Read March 10, 1797.

IN a letter of July 3d, I informed our late moſt worthy preſident that ſome bones of a very large animal of the clawed kind had been recently diſcovered within this ſtate, and promiſed a communication on the ſubject as ſoon as we could recover what were ſtill recoverable of them. It is well known that the ſubſtratum of the country beyond the Blue Ridge is a limeſtone, abounding with large caverns, the earthy floors of which are highly impregnated with nitre ; and that the inhabitants are in the habit of extracting the nitre from them. In digging the floor of one of theſe caves, belonging to Frederic Cromer in the county of Greenbriar, the labourers at the depth of two or three feet, came to ſome bones, the ſize and form of which beſpoke

bespoke an animal unknown to them. The nitrous impregnation of the earth together with a small degree of petrification had probably been the means of their preservation. The importance of the discovery was not known to those who made it, yet it excited conversation in the neighbourhood, and led persons of vague curiosity to seek and take away the bones. It was fortunate for science that one of its zealous and well informed friends, Colonel John Stewart of that neighbourhood, heard of the discovery, and, sensible from their description, that they were of an animal not known, took measures without delay for saving those which still remained. He was kind enough to inform me of the incident, and to forward me the bones from time to time as they were recovered. To these I was enabled accidentally to add some others by the kindness of a Mr. Hopkins of New-York, who had visited the cave. These bones are,

1*st*. A small fragment of the femur or thigh bone; being in fact only its lower extremity, separated from the main bone at its epiphysis, so as to give us only the two condyles, but these are nearly entire.

2*d*. A radius, perfect.

3*d*. An ulna, or fore-arm, perfect, except that it is broken in two.

4*th*. Three claws, and half a dozen other bones of the foot; but whether of a fore or hinder foot, is not evident.

About a foot in length of the residue of the femur was found, it was split through the middle, and in that state was used as a support for one of the salt petre vats, this piece was afterwards lost, but its measures had been first taken as will be stated hereafter.

These bones only enable us to class the animal with the unquiculated quadrupeds; and of these the lion being nearest to him in size, we will compare him with that animal, of whose anatomy Monsieur Daubenton has furnished very accurate measures in his tables at the end of Buffon's Natural History of the lion. These measures were taken as he* informs us from " a large lion of Africa," in which quarter the largest † are said to be produced. I shall select from his measures only those where we have the corresponding bones, converting them into our own inch and its fractions, that the comparison may be more obvious: and to avoid the embarrassment of designating our animal always by circumlocution and description, I will venture to refer to him by the name of the Great-Claw or Megalonyx, to which he seems sufficiently entitled by the distinguished size of that member.

	Megalonyx. Inches	Lion. Inches.
Length of the ulna, or fore-arm	20.1	13.7
Height of the olecranum	3.5	1.85
Breadth of the ulna, from the point of the coronoide apophysis to the extremity of the olecranum	9.55	
Breadth of the ulna at its middle	3.8	
Thickness at the same place	1.14	
Circumference at the same place	6.7	
Length of the radius	17.75	12.37
Breadth of the radius at its head	2.65	1.38
Circumference at its middle	7.4	3.62
Breadth at its lower extremity	4.05	1.18
Diameter of the lower extremity of the femur at the base of the two condyles	4.2	2.65
Transverse diameter of the larger condyle at its base	3.	
Circumference of both condyles at their base	11.65	

* Buffon, XVIII. 38. Paris edition in 31 vols. 12mo.
† 2. De Manet, 117.

Diameter

CERTAIN BONES, &c.

	Megalonyx. Inches.	Lion. Inches.
Diameter of the middle of the femur	4.25	1.15
Hollow of the femur at the same place	1.25	
Thickness of the bone surrounding the hollow	1.5	
Length of the longest claw	7.5	1.41
Length of the second phalanx of the same	3.2	1.41

The dimensions of the largest of the foot bones are as follow,

	Inches.
Its greatest diameter, or breadth at the joint	2.45
Its smallest diameter, or thickness at the same place	2.28
Its circumference at the same place	7.1
Its circumference at the middle	5.3

	Of longest toe.	Middle sized toe.	Shortest toe.
2d. Phalanx. Its length	3.2	2.95	
Greatest diameter at its head or upper joint	1.84	2.05	
Smallest diameter at the same place	1.4	1.54	
Circumference at the same place	5.25	5.8	
3d. Phalanx. Its length	*7.5	†5.9	3.5
Greatest diameter at its head or upper joint	2.7	2.	1.45
Smallest diameter at the same place	.95	.9	.55
Circumference at the same place	6.45	4.8	

Were we to estimate the size of our animal by a comparison with that of the lion on the principle of *ex pede Herculem*, by taking the longest claw of each as the mo-

---

* It is actually 6¾ inches long, but about ¾ inch appear to have been broken off.

† Actually 5.65 but about ¼ inch is broken off.

dule

dule of their measure, it would give us a being out of the limits of nature. It is fortunate therefore that we have some of the larger bones of the limbs which may furnish a more certain estimate of his stature. Let us suppose then that his dimensions of height, length and thickness, and of the principal members composing these, were of the same proportions with those of the lion. In the table of M. Daubenton an ulna of 13.78 inches belonged to a lion $42\frac{1}{2}$ inches high over the shoulders: then an ulna of 20.1 inches bespeaks a megalonyx of 5 feet 1.75 inches height, and as animals who have the same proportions of height, length, and thickness have their bulk or weights proportioned to the cubes* of any one of their dimensions, the cube of 42.5 inches is to 262 lb. the height and weight of M. Daubenton's lion as the cube of 61.75 inches to 803 lb. the height and weight of the magalonyx; which would prove him a little more than three times the size of the lion. I suppose that we should be safe in considering, on the authority of M. Daubenton, his lion as a large one. But let it pass as one only of the ordinary size, and that the megalonyx whose bones happen to have been found was also of the ordinary size. It does † appear that there was dissected for the academy of sciences at Paris, a lion of 4 feet $9\frac{3}{8}$ inches height. This individual would weigh 644 lb. and would be in his species, what a man of eight feet height would be in ours. Such men have existed. A megalonyx equally monstrous would be 7 feet high, and would weigh 2000 lb. but the ordinary race, and not the monsters of it, are the object of our present enquiry.

I have used the height alone of this animal to deduce his bulk, on the supposition that he might have been formed in the proportions of the lion. But these were

---

* Buffon xxii. 121. † Buffon xviii. 15.

not his proportions, he was much thicker than the lion in proportion to his height, in his limbs certainly, and probably therefore in his body. The diameter of his radius, at its upper end, is near twice as great as that of the lion, and, at its lower end, more than thrice as great, which gives a mean proportion of $2\frac{1}{2}$ for 1. The femur of the lion was lefs than $1\frac{1}{4}$ inch diameter. That of the megalonyx is $4\frac{1}{4}$ inches, which is more than three for one. And as bodies of the fame length and fubftance have their weights proportioned to the fquares of their diameters, this excefs of caliber compounded with the height, would greatly aggravate the bulk of this animal. But when our fubject has already carried us beyond the limits of nature hitherto known, it is fafeft to ftop at the moft moderate conclufions, and not to follow appearances through all the conjectures they would furnifh, but leave thefe to be corroborated or corrected by future difcoveries. Let us only fay then, what we may fafely fay, that he was *more* than three times as large as the lion: that he ftood as pre-eminently at the head of the column of clawed animals as the mammoth ftood at that of the elephant, rhinoceros, and hippopotamus: and that he may have been as formidable an antagonift to the mammoth as the lion to the elephant.

A difficult queftion now prefents itfelf. What is become of the great-claw? Some light may be thrown on this by afking another queftion. Do the wild animals of the firft magnitude in any inftance fix their dwellings in a thickly inhabited country? fuch, I mean, as the elephant, the rhinoceros, the lion, the tyger? as far as my reading and recollection ferve me, I think they do not: but I hazard the opinion doubtingly, becaufe it is not the refult of full enquiry. Africa is chiefly inhabited along the margin of its feas and rivers. The interior defart is the domain of the elephant, the rhinoceros, the lion,

lion, the tyger. Such individuals as have their haunts nearest the inhabited frontier, enter it occasionally, and commit depredations when preffed by hunger: but the mafs of their nation (if I may ufe the term) never approach the habitation of man, nor are within reach of it. When our anceftors arrived here, the Indian population, below the falls of the rivers, was about the twentieth part of what it now is. In this ftate of things, an animal refembling the lion feems to have been known even in the lower country. Moft of the accounts given by the earlier adventurers to this part of America make a lion one of the animals of our forefts. Sir John Hawkins * mentions this in 1564. Thomas Harriot, a man of learning, and of diftinguifhed candor, who refided in Virginia in 1587 † does the fame, fo alfo does Bullock in his account of Virginia,‡ written about 1627, he fays he drew his information from Pierce, Willoughby, Claiborne, and others who had been here, and from his own father who had lived here twelve years. It does not appear whether the fact is ftated on their own view, or on information from the Indians, probably the latter. The progrefs of the new population would foon drive off the larger animals, and the largeft firft. In the prefent interior of our continent there is furely fpace and range enough for elephants and lions, if in that climate they could fubfift; and for mammoths and megalonyxes who may fubfift there. Our entire ignorance of the immenfe country to the Weft and North-Weft, and of its contents, does not authorife us to fay what it does not contain.

Moreover it is a fact well known, and always fufceptible of verification, that on a rock on the bank of the

---

* Hakluyt, 541. edition of 1589.
† Ibid. 757, and Smith's Hiftory of Virginia, 10.
‡ Bullock, page 5.

Kanhawa, near its confluence with the Ohio, there are carvings of many animals of that country, and among these one which has always been confidered as a perfect figure of a lion. And these are so rudely done as to leave no room to fufpect a foreign hand. This could not have been of the fmaller and manelefs lion of Mexico and Peru, known also in Africa both in * ancient and † modern times, though denied by ‡ M. de Buffon: becaufe like the greater African lion, he is a tropical animal; and his want of a mane would not fatisfy the figure. This figure then muft have been taken from some other prototype, and that prototype muft have refembled the lion fufficiently to fatisfy the figure, and was probably the animal the defcription of which by the Indians made Hawkins, Harriot, and others conclude there were lions here. May we not prefume that prototype to have been the great-claw?

Many traditions are in poffeffion of our upper inhabitants, which themfelves have heretofore confidered as fables, but which have regained credit fince the difcovery of thefe bones. There has always been a ftory current that the firft company of adventurers who went to feek an eftablifhment in the county of Greenbriar, the night of their arrival were alarmed at their camp by the terrible roarings of fome animal unknown to them: that he went round and round their camp, that at times they faw his eyes like two balls of fire, that their horfes were fo agonifed with fear that they couched down on the earth, and their dogs crept in among them, not daring to bark. Their fires, it was thought, protected them, and the next morning they abandoned the country. This was little more than 30 years ago.—In the year 1765, George Wilfon and John Davies, having gone to hunt

---

* Ariftot. Animal, 9. 4. Pliny, 8. 16.   † Kolbe.   ‡ Buffon, xviii. 18.

on Cheat river, a branch of the Monongahela, heard one night, at a distance from their camp, a tremendous roaring, which became louder and louder as it approached, till they thought it resembled thunder, and even made the earth tremble under them. The animal prowled round their camp a considerable time, during which their dogs, though on all other occasions fierce, crept to their feet, could not be excited from their camp, nor even encouraged to bark. About day light they heard the same sound repeated from the knob of a mountain about a mile off, and within a minute it was answered by a similar voice from a neighbouring knob. Colonel John Stewart had this account from Wilson in the year 1769, who was afterwards Lieutenant Colonel of a Pennsylvania regiment in the revolution-war; and some years after from Davies, who is now living in Kentucky.

These circumstances multiply the points of resemblance between this animal and the lion. M. de la Harpe of the French Academy, in his abridgment of the General History of Voyages, speaking of the Moors, says* " it is remarkable that when, during their huntings, they meet with lions, their horses, though famous for swiftness, are siezed with such terror that they become motionless, and their dogs equally frightened, creep to the feet of their master, or of his horse." Mr. Sparrman in his voyage to the Cape of Good Hope, chap. 11. says, " we could plainly discover by our animals when the lions, whether they roared or not, were observing us at a small distance. For in that case the hounds did not venture to bark, but crept quite close to the Hottentots; and our oxen and horses sighed deeply, frequently hanging back, and pulling slowly with all their might at the strong straps with which they were tied to the waggon. They

* Gentleman's, and London Magazines, for 1783.

also

alſo laid themſelves down on the ground, and ſtood up alternately, as if they did not know what to do with themſelves, and even as if they were in the agonies of death." He adds that " when the lion roars, he puts his mouth to the ground, ſo that the ſound is equally diffuſed to every quarter." M. de Buffon (xviii. 31.) deſcribes the roaring of the lion as, by its echoes reſembling thunder: and Sparrman c. 12. mentions that the eyes of the lion can be ſeen a conſiderable diſtance in the dark, and that the Hottentots watch for his eyes for their government. The phoſphoric appearance of the eye in the dark ſeems common to all animals of the cat kind.

The terror excited by theſe animals is not confined to brutes alone. A perſon of the name of Draper had gone in the year 1770, to hunt on the Kanhawa. He had turned his horſe looſe with a bell on, and had not yet got out of hearing when his attention was recalled by the rapid ringing of the bell. Suſpecting that Indians might be attempting to take off his horſe, he immediately returned to him, but before he arrived he was half eaten up. His dog ſcenting the trace of a wild beaſt, he followed him on it, and ſoon came in ſight of an animal of ſuch enormous ſize, that though one of our moſt daring hunters and beſt markſmen, he withdrew inſtantly, and as ſilently as poſſible, checking and bringing off his dog. He could recollect no more of the animal than his terrific bulk, and that his general outlines were thoſe of the cat kind. He was familiar with our animal miſcalled the panther, with our wolves and wild beaſts generally, and would not have miſtaken nor ſhrunk from them.

In fine, the bones exiſt: therefore the animal has exiſted. The movements of nature are in a never ending circle. The animal ſpecies which has once been put into a train of motion, is ſtill probably moving in that train. For if one link in nature's chain might be loſt, another and another might be loft, till this whole fyftem of things fhould evanifh by piece-meal; a conclufion not warranted by the local difappearance of one or two fpecies of animals, and oppofed by the thoufands and thoufands of inftances of the renovating power conftantly exercifed by nature for the reproduction of all her fubjects, animal, vegetable, and mineral. If this animal then has once exifted, it is probable on this general view of the movements of nature that he ftill exifts, and rendered ftill more probable by the relations of honeft men applicable to him and to him alone. It would indeed be but conformable to the ordinary economy of nature to conjecture that fhe had oppofed fufficient barriers to the too great multiplication of fo powerful a deftroyer. If lions and tygers multiplied as rabbits do, or eagles as pigeons, all other animal nature would have been long ago deftroyed, and themfelves would have ultimately extinguifhed after eating out their pafture. It is probable then that the great-claw has at all times been the rareft of animals. Hence fo little is known, and fo little remains of him. His exiftence however being at length difcovered, enquiry will be excited, and further information of him will probably be obtained.

The Cofmogony of M. de Buffon fuppofes that the earth and all the other planets primary and fecondary, have been maffes of melted matter ftruck off from the fun by the incidence of a comet on it: that thefe have been cooling by degrees, firft at the poles, and afterwards more and more towards their Equators: confequently that on our earth there has been a time when the temperature of the poles fuited the conftitution of the elephant, the rhinoceros, and hippopotamus: and in proportion as the remoter zones became fucceffively too cold, thefe animals have retired more and more towards the Equatorial regions, till now that they are reduced to the

the torrid zone as the ultimate stage of their existence. To support this theory, he *assumes the tusks of the mammoth to have been those of an elephant, some of his teeth to have belonged to the hippopotamus, and his largest grinders to an animal much greater than either, and to have been deposited on the Missouri, the Ohio, the Holston, when those latitudes were not yet too cold for the constitutions of these animals. Should the bones of our animal, which may hereafter be found, differ only in size from those of the lion, they may on this hypothesis be claimed for the lion, now also reduced to the torrid zone, and its vicinities, and may be considered as an additional proof of this system; and that there has been a time when our latitudes suited the lion as well as the other animals of that temperament. This is not the place to discuss theories of the earth, nor to question the gratuitous allotment to different animals of teeth not differing in any circumstance. But let us for a moment grant this with his former postulata, and ask how they will consist with another theory of his " qu'il y a dans la combinaison des elemens et des autres causes physiques, quelque chose de contraire a l'aggrandisement de la nature vivante *dans ce nouveau monde;* qu'il y a des obstacles au developpement et peutetre a la formation des grands germes †." He says that the mammoth was an elephant, yet ‡ two or three times as large as the elephants of Asia and Africa: that some of his teeth were those of a hippopotamus, yet of a hippopotamus § four times as large as those of Africa: that the mammoth himself, for he still considers him as a distinct animal,‖ " was of a size superior to that of the largest elephants. That he was the primary and greatest of all terrestrial

---

* Buffon, Epoq. 2. 233, 234. † Buffon, xviii. 145. ‡ 2. Epoq. 223.
§ 1. Epoq. 246. 2. Epoq. 232.   ‖ 2. Epoq. 234, 235.

animals."

animals." If the bones of the megalonyx be afcribed to the lion, they muft certainly have been of a lion of more than three times the volume of the African. I delivered to M. de Buffon the fkeleton of our palmated elk, called orignal or moofe, 7 feet high over the fhoulders, he is often confiderably higher. I cannot find that the European elk is more than two thirds of that height: confequently not one third of the bulk of the American. He* acknowledges the palmated deer (daim) of America to be larger and ftronger than that of the Old World. He † confiders the round horned deer of thefe States and of Louifiana as the roe, and admits they are of three times his fize. Are we then from all this to draw a conclufion, the reverfe of that of M. de Buffon. That nature, has formed the larger animals of America, like its lakes, its rivers, and mountains, on a greater and prouder fcale than in the other hemifphere? Not at all, we are to conclude that fhe has formed fome things large and fome things fmall, on both fides of the earth for reafons which fhe has not enabled us to penetrate; and that we ought not to fhut our eyes upon one half of her facts, and build fyftems on the other half.

To return to our great-claw; I depofit his bones with the Philofophical Society, as well in evidence of their exiftence and of their dimenfions, as for their fafe-keeping; and I fhall think it my duty to do the fame by fuch others as I may be fortunate enough to obtain the recovery of hereafter.

TH: JEFFERSON.

Monticello, Feb. 10*th*, 1797.

* Buffon, xxix. 245.   † Ibid. xii. 91. 92. xxix. 245. Vide Suppl. 201.

[*Editor's Note:* A postscript to the memoir (pp. 259-260) has been omitted for the sake of brevity. It treats the resemblance between Jefferson's *Megalonyx* and a recently discovered fossil "Megatherium" from Paraguay. In the final article of the *American Philosophical Society Transactions* volume 4, 1799, Caspar Wistar published detailed descriptions of Jefferson's fossils, with two plates. Wistar concluded that the Virginia bones were distinct from those of the Paraguay discovery.]

FACTS, OBSERVATIONS, AND CONJECTURES, RELATIVE TO THE ELEPHANTINE BONES (OF DIFFERENT SPECIES), THAT ARE FOUND IN VARIOUS PARTS OF NORTH AMERICA

Benjamin Smith Barton

DEAR SIR,

I KNOW that every new discovery in Natural History will give you pleasure. But I am persuaded, that your pleasure is always *peculiarly* great, when such discoveries tend to throw any light upon the curious subject of *extinct* species of animals: a subject to which you have devoted so much, and such successful, attention.

Without further delay, I hasten to inform you of a recent discovery relative to the Mammoth*, or American Elephant. If the facts be as I state them, I think you will not hesitate to consider the discovery one of the most interesting that has been made for a long time. I may add, that such a discovery was hardly to be expected, by the most sanguine or enthusiastic zoologist.

Very lately, in digging a well, near a salt-lick, in the county of Wythe, in Virginia, after penetrating about five feet and a half below the surface of the soil, the workmen struck upon the *stomach* of one of those huge animals, best known, in the United-States, by the name of the Mammoth. The contents of the viscus were carefully examined, and were found to be " in a state of perfect preservation." They consisted of half-masticated reeds (a species of Arundo, or Arundinaria, still common in Virginia, and other parts of the United-States), of twigs of trees, and of grass, or leaves.— " There could (says my informant) be no deception on the subject. The substances were designated by obvious characters, which could not be mistaken, and of which every one could judge: besides, the bones of the

---

* In compliance with the usage of my countrymen, I call this animal Mammoth, or Mammouth, though I well know, that this appellation is more properly bestowed upon another species of Elephant, the remains of which are very numerous in various parts of Asia: the Elephas primigenius of my excellent friend Professor Blumenbach, and your Elephas Mammonteus.

animal lay around, and added a silent, but sure, confirmation."

All the vestiges, which I have mentioned, were incumbent upon a stratum of limestone. From the number of bones already discovered, hopes are entertained, that a complete skeleton of this enormous animal, once so common in many parts of the New-World, may be formed.

The information, which I have communicated to you, I have just received from one of my correspondents in Virginia, Bishop Madison, the President of the College of William and Mary, in that state. The Bishop is a man of considerable attainments in science, and has long enjoyed the reputation of being one of the most amiable and respectable characters in our country. Mr. Madison's letter is dated (Williamsburg) October 6th; so that, you see, I have not, in this instance at least, lost much time in letting you know what we are doing, in the United-States, for your favourite study.

It may be proper to inform you, that the county of Wythe, in which the exuviæ were discovered, is one of the trans-alpine counties of the state: that is, it is situated to the west of the great ranges of mountains known, in our country, by the names of the Blue-Ridge (or South-Mountain*); the North-Mountain; and the Alleghaney-Mountain; and the spot may be about 200,

* Sometimes called the Blue-Mountain. This great chain has been confounded, by Professor Playfair, and other learned naturalists, with the Alleghaney-Mountain.

or 250, miles from the *nearest* part of the Atlantic-Ocean. This county borders on that of Greenbryar, in which the bones of the Megalonyx, as Mr. Jefferson has denominated it, were discovered. I will only further add, that, in the view of the naturalist, this is one of the most interesting portions of Virginia, and, perhaps, of North-America. The floor of the country is limestone (different varieties of psadurium), rich in the impressions of numerous species of sea *testacea*, and other marine animals: the caverns abound in nitre, and in sulphates of soda and magnesia; while springs of various gaseous and mineral impregnations, and of different temperatures, present themselves almost every where. I say nothing of the manganese, and various other metallic bodies, which have lately been detected*.

I shall not take up any of your time in endeavouring to prove, that the *soft* parts of animal bodies, such as the skin, the muscles, the stomach, &c., may be preserved, in a state of considerable perfection, for a great length of time. You, Sir, are well acquainted with the various facts, relative to this subject, that have been published by Dr. Pallas, and some other eminent naturalists. I will, however, take the liberty of referring you to the first part of my *Medical and Physical Journal*, pages 154—159, for a very interesting notice concerning the discovery of five skeletons of Mammoths, near the river Ohio. From this account, it would seem pretty cer-

---

* In the same tract of country, large quantities of sulphate of barytes have been found.

tain, that so late as 1762, which was, in all probability, several centuries after the extinction of the species in America, the proboscis *(trompe)* of one of the animals was preserved: for the Indians, in their account of the discovery, said, that the head of one of the Mammoths was furnished " with a long nose, and the mouth on the under side." This long nose, I have no doubt, was the proboscis. Since the publication of the notice, to which I have referred you, I have observed, in Kalm's *Travels*, a circumstance which deserves to be repeated here. Speaking of an enormous skeleton, supposed to be that of an Elephant, which was found by the Indians in a swamp, " in that part of Canada where the Illinois live," the honest Swedish traveller says, that he was informed by an officer, who had seen the remains, " that the figure of the whole snout was still clearly visible, though it was now half mouldered*." The snout, as it is here called, seems to refer to the proboscis, or trompe. Indeed, the Swedish word " *snabelen*," in the original, leaves us in little doubt on the subject.

I have no reason to believe, that the skeleton, of which Kalm speaks, was one of those of which mention is made in my *Journal*. The contrary is more probable. Be this as it may, it would appear, from the double testimony which I have collected on the subject, that not only the bones, but even the long nose, or proboscis, of the American elephant, has been preserved, and seen, in some of the marshes of the country.

* En Resa til Norra America, &c., af Pehr Kalm. Tom. III. p. 244. See, also, the English translation, by Dr. J. R. Forster. Vol. III. p. 11 and 12. London: 1771.

The salt-licks, or marshes, in which so many of the bones of the Mammoth have been found, seem very well adapted for the preservation of both the hard and soft parts of animal bodies. Some of these licks are *muriatic* marshes, or marshes impregnated with muriate of soda, and even at this day abound in Salicornia, Glaux, Triglochin, and other plants, which are rarely found at any great distance from such saline soils, which in America, as in other countries, doubtless, owe their origin to the sea. Other North-American salt-licks seem more impregnated with sulphate of magnesia, or epsom, than with muriatic, salt; while others of them, again, are very sensibly impregnated with sulphate of alumine, or with sulphate of iron. Lastly, some of the licks seem to be very little different from your sphagnum morasses in Europe, in some of which, it is well known to you, that the bones of a species of Cervus (allied to the Alces), and those of other animals, have been preserved, for a very great length of time. (See Note 1, at the end of this article). You will observe, Sir, that the Wythe exuviæ, recently discovered, were found " near a salt-lick;" and it is probable, that when we shall receive a more circumstantial account of the discovery, it will clearly appear, that the stomach, bones, &c., were exposed to the influence of the saline impregnation; and that it is to this that we are, in a considerable degree, indebted for their preservation: a preservation so precious to the lovers of Natural History.

We shall never, perhaps, be certain at what period the *species* of the Mammoth ceased to exist in America. We may, however, I think, confidently assert, that several centuries have elapsed since this vast animal was a *common* inhabitant of the forests or marshes of this continent; for none of the earliest visitors of America (if we except some idle travellers, by no means studious of the truth) pretend to have seen a quadruped, in any respect, allied to the elephant of the New-World. (See Note 2.) Neither do I learn, that they received, from the native inhabitants, any traditional information relative to the *recent* existence of such an animal. Now more than three centuries have elapsed since the discovery of the New-World by Columbus and Vespucci. Above two centuries and a half have elapsed since Spanish armies, in pursuit of gold, rambled over immense portions of the country now called Georgia, and over the two Floridas, on both sides of the Missisippi; and it is almost two centuries since the English first visited Virginia, and even founded colonies in that country. Nor were the visits of the English, at this early period, confined to the maritime, or most eastern, part of the country. They often penetrated as far as the first and second ranges of mountains, and explored those very tracts of country in which the bones of the Mammoth (as well as those of the Megatherium) have been recently found. But they saw no living representatives of the vestiges of either of these animals. Upon the whole, I think we proceed upon a pretty solid foundation when we assert, that almost the entire race of the Mammoth has been extinct for much more than three hundred years. It is, indeed, highly probable, that a few *individuals* of the species may have existed for many years, perhaps a century, or double this term of time, after *the greater part* of the species had disappeared. It is even possible, but not, I think, very probable, that a few solitary Mammoths may have trod the country to the east of the Missisippi, *since* the first discovery of the continent of North-America. Perhaps, those of which the proboscides, the stomachs, and other soft parts, have been preserved, were some of the *last-surviving individuals* of this stupendous species, which Nature (for purposes unknown to us) has removed from the number of *living* existences.

The chief value of the recent discovery, in Virginia, seems to consist in the ascertaining of this fact, that the Mammoth was an herbivorous, and not a carnivorous, animal. The discovery " has summoned (to use Bishop Madison's words) the discordant opinions of philosophers before a tribunal, from which there is no appeal."

As to myself, I have always leaned to the opinion, that the Mammoth was an herbivorous animal. I have even, for at least six years, defended this opinion, in my public lectures; as I have, also, the opinion, that the Mammoth was a species of Elephas*. In respect to the first opinion, I was well aware, that I had not a few respectable authorities to oppose. Among these, there were some ingenious countrymen of my own; and among the foreigners, not to mention others, the late Mr. John Hunter. In a conversation which I had with that truly ingenious man, in the year 1787, on the subject of the Mammoth, he observed to me, in a style

---

* See my letter to Mons. Lacépède, in Mr. Tilloch's Philosophical Magazine, for July, 1805.

rather authoritative, "that the *Incognitum* had, certainly, been a carnivorous animal." You know, Sir, that the same opinion had been entertained, and given to the public, by Mr. Hunter's brother, the celebrated Dr. William Hunter*, almost twenty years before the period I have mentioned.

North-America appears to have been the favourite, but not (I think) the exclusive, domain of the Mammoth. The exuviæ of this giant of the earth have been found in almost every state of the American Union. They have been discovered in the countries west of the Mississippi, as well as in those which are included between this river and the Atlantic-Ocean. Consequently, the Alleghaney-Mountains, the Blue-Ridge, and other ranges of our mountains, formed but a feeble barrier against his passage from the west to the east, or from the east to the west. The medals of his existence remain in a thousand places; and in a few years, I trust, we shall be able to speak, with some degree of certainty, concerning the extent of his geographical range through the continent. At present, I do not recollect any proofs of his existence in a higher latitude than 43°†. But I am far from supposing, that the Mammoth ceased to exist to the north of this degree. When Mr. Jefferson wrote his *Notes*, he was unable to trace this species of elephant

---

* See his Observations on the Bones, commonly supposed to be Elephant's Bones, which have been found near the River Ohio, in America, in the Philosophical Transactions, vol. 58, for the year 1768.

† In the neighbourhood of Lake-Erie.

to the south of lat. 36½°, in the tract of country now called Tennessee*. But we are now well assured, that this quadruped had existed, in many parts of America, several degrees below the most southern limits of the state of Tennessee.

I have said, that the ranges of our mountains did not prevent the passage of the Mammoth from the western to the eastern, or from the eastern to the western, parts of the continent. I am much inclined, however, to believe, that this quadruped has *always* been a much more common animal in the countries to the west, than in those to the east, of the Alleghaney-Mountains. Certain it is, that we have already discovered a much greater number of the Mammoth's remains in the former than in the latter of these districts; although, from the progress of settlement in, and from the explorations of, the continent of North-America, the very reverse should have been expected, admitting it to be a fact, that the remains were equally abundant, in an equal extent of country, on both sides of the mountains. I need say nothing to you concerning the immense collections of Mammoth's bones that have, at various periods, been discovered in Kentucky, particularly in and about the great salt-licks. Collections, not less extensive, have been discovered to the west of the Missisippi. But in the tract of country to the east of the Alleghaney and North-Mountains, we not only have not discovered these vestiges so abundantly cumulated, but we have disco-

---

* Notes on the State of Virginia; written in the year 1781, &c. pages 71, 76. Original edition, printed in 1782.

vered them in a much smaller number of places. It must not be concealed, however, that Mr. Peale's two skeletons were found in the latter tract of country; and some of the bones of a Mammoth have been discovered in the state of New-Jersey, at the distance of a few miles from Philadelphia.

It is, perhaps, worth observing, in this place, that the different kinds of licks, especially the muriatic marshes, and transparent springs of water impregnated with muriate of soda, are much more commonly met with in the western than in the eastern parts of North-America. I do not mention this as a *decided* confirmation of my position, that the Mammoth was more common in the trans-alpine, than in the Atlantic, or submaritime, parts of the continent: for it is, certainly, possible, that we may have discovered more of this elephant's exuviæ in the western than in the eastern countries *chiefly* because they were more likely to be preserved (owing to the greater number of marshes) in the former than in the latter countries. But I cannot help suspecting, that the Mammoth, like the bison, the elk, and the other animals formerly enumerated, resorted to the licks, for the purpose of *drinking* the saline water, and of *eating* the earth impregnated with it.

If future and more extensive researches should more clearly establish my position, relative to the diffusion of the Mammoth across the continent of North-America, it will be somewhat remarkable, that the bones of this animal have so seldom been seen in the eastern parts of

Asia, from whence I have no doubt, that many of the animals of America have been derived.

I am far, however, from supposing, that Asia has been the parental country of *all* the animals that have been found in the two continents and islands of the New-World*. But I have observed, and it is a circumstance much in favour of the hypothesis which considers Asia as the fountain from whence have proceeded *many* of the American animals, that where the same species of quadruped is common to these two portions of the earth, they are *generally* more common in the western than in the eastern districts of America. This rule is, perhaps, liable to *some* exceptions: but this is chiefly the case when the quadruped is found in Europe, as well as in Asia and America. Thus, I cannot assert, that the Beaver (Castor Fiber) is more common in the western than it is in the eastern parts of North-America. But, then, the Beaver, it will be recollected, is one of those quadrupeds which are common to Asia, to Europe, and to America.

Does not this fortunate Virginia discovery give us pretty good reason to believe, that at some future, and perhaps not distant, period, the labours of workmen, intent upon very different objects, will exhibit to us the Mammoth in a state not less perfect than that in which Pallas had an opportunity of contemplating the Rhinoceros, near the banks of the river Willioni, in the north of Asia? Let us cease, then, to deplore the inherent imperfections

* See my New Views of the Origin of the Tribes and Nations of America. Preliminary Discourse, pages ci, cii, ciii.

of zoological science on the score of *lost* species. Doubtless, *many species of animals have ceased to exist*. But of not a few of these we shall be able to ascertain the precise forms and characters (and even the *mores*, or manners); and thus, Sir, to assign to each its proper place, in that more finished view of the animal creation, for which your laborious researches, the researches of many of your countrymen, and of the learned in other parts of the world, are rapidly preparing us.

I fear I have fatigued you. But read on, to be assured of the high esteem with which I am,
Dear Sir,
Your friend, &c.
BENJAMIN SMITH BARTON.
Philadelphia, October 14th,
1805.

NOTES ON THE PRECEDING PAPER.

Note 1. *Page* 27. The licks, of which I have given some account, are resorted to by various species of animals, particularly Deer (Cervus virginianus), the American Elk (my Cervus Wapiti), and the Bison, or Bos americanus of Gmelin. It is a fact, not generally known, that the animals which I have mentioned not only *lick* the soil, and *drink* the water of these salines, but even *chew* and *swallow* the ground, in large quantities. Hence, our Indians designate some of the licks by a name which may be translated

" the chewing place."——Besides the animals already mentioned, horses are observed to be very fond of drinking the water of some of the licks. But, which is more remarkable, they are frequented by vast numbers of Wild Pigeons (Columba migratoria), and by a species of Psittacus. Some species of Crotalus, or Rattle-Snake, are also often observed about these places: but whether these reptiles drink the saline water, I cannot assert.——So far as I have yet learned, it is the *herbivorous* mammalia only that resort to the licks, for the purpose of drinking the water, and of licking and eating the earth. This circumstance ought not to have been overlooked, in the view of the question, whether the American Mammoth was an herbivorous or a carnivorous animal. By myself, indeed, it was not overlooked.

Note 2. *Page* 28. One David Ingram, an Englishman, assures us, that he saw Elephants in America; and we might, perhaps, repose *some* degree of confidence in his assertion, if he did not tell us, that he likewise saw wild animals, twice the size of our horses, formed like a grey-hound in their hinder parts; another quadruped, larger than the bear, without head or neck, having its eyes and mouth in its breast; and, lastly, the DEVIL, sometimes in the likeness of a dog, at other times in that of a calf!

6

ACCOUNT OF FOSSIL SHELLS, WITH THE AUTHOR'S REASONS FOR ATTENDING TO THE SAME

In a Letter to Levi Hedge, F. A. A.

Parker Cleaveland
Professor of Mathematics and Natural Philosophy in Bowdoin College

———

Bowdoin College, 10 *October*, 1808.

DEAR SIR,

AT your request I transmit you an account of the fossil shells, which you saw in my possession, when I had the pleasure of your visit at Brunswick. Previous however to a relation of the particulars, I will take the liberty of mentioning the reasons, which have induced me to pay attention to facts of so common occurrence.

The universal existence of marine shells and other fossil bodies, at considerable depths below the surface of the earth, satisfactorily prove that very great changes have taken place in the exterior parts of our globe, either by sudden and powerful convulsions, or in some more gradual manner. In every system of geology fossil bodies have deservedly received a large share of attention; and it is perhaps true, that further discoveries of fossil shells on mountains and in very elevated situations, under circumstances precisely similar to those, in which they have been found, would afford very little assistance in forming more correct geological systems. But with regard to the discovery of shells in plains and small elevations near the sea, the preceding remark may not be true. Concerning these it may be inquired, whether the changes, to which the shells owe their existence as fossils, be of ancient or modern date; whether they were produced by great convulsions and sudden inundations, or by gradual alterations of many successive years. It has been suggested that important advantages would result from possessing a geographical map, indicating the different species of fossil shells, and the places, in which they were found. I think the idea important, and practicable at least with regard to any country or coast, which may be thickly inhabited. With such a map before us we should be better enabled to compare individual facts, and hence to draw several conclusions. Under this view of the subject the discovery of shells, which are merely fluviatile, will be worthy of attention. For the preceding reasons I have endeavoured to collect all the facts in my power. I will now give you a description of two wells, which I have examined the last summer, while digging.

One is in Bowdoin, at the distance of three or four miles from the nearest salt water, which is at the termination of the tide in Cathance river; the distance of the well from the sea is probably about twenty miles. Its elevation above the tide in Cathance river is estimated by gentlemen, living in that part of the country, at seventy or eighty feet. The land about the well is very uneven, and abounds with gneiss and a coarse granite. A small stream passes about fifty rods from the well; and, after a long and winding course, assists in forming the Cathance. This well is twenty feet deep. Through the first ten feet from the surface a hard gravel is found, stratified and interspersed with layers of coarse, yellowish sand. At the depth of ten feet commences a stratum of blue clay, into which the workmen dug ten feet, but without passing through the clay. When first taken from the ground, it is nearly black, and very tenacious. This clay both in appearance and smell resembles that dug on flats, or near salt marshes, or on the margin of salt water rivers. The shells also have the same smell, when first taken from the clay; and, as far as I have seen, are the clam, and two varieties of the muscle; and another kind of shell, whose genus I know not. It is large, of a conical form, about three inches in length, and passing in a double spiral line from the larger part to the vertex. The same genus is found on our sea shores. These shells are in general well preserved, and in almost every instance filled with clay; which must have entered them with the water, in which it was suspended. I saw very few valves lying by themselves. When carefully taken from the clay, the shell is either whole, or the valves opened and lying contiguous to each other at the hinge. I also took from near the bottom of the same well a large rock, to which were adhering many of those shells, which the seamen call barnacles.

The other well is situated in Brunswick; at an elevation of about eighty feet above the tide water in the Androscoggin; and about half a mile westward of the same river above the falls. It is on the side of a hill. Several gullies take their rise at the foot of the hill, and lead to the river. This well is twenty two feet deep. After cutting through the soil, the first twelve feet consist of alternate strata of yellowish sand and common brick clay. At the depth of twelve feet commences a stratum of blue clay, which is four feet thick. This clay is plentifully interspersed with shells, similar to those before mentioned; and has the same appearance and smell. This is followed by a stratum of grey sand, similar to that often seen upon beaches. The next three feet consist of thin, alternate layers of common brick clay and a reddish sand. The last stratum, and that, in which the well terminates, is a brown sand, resembling that, which is frequently found at the surface.

I have selected these two wells from several others, because I had better opportunities of examining them.

I have a few specimens, which I should be happy in sending you, were the opportunity convenient.

I am, dear sir,
you friend &c.
PARKER CLEAVELAND.

OBSERVATIONS ON A NEW GENUS OF FOSSIL SHELLS

Charles Alexander Le Sueur

THE secondary blue limestone of which the great basin is composed, which extends from the Alleghany Mountains to Lake Superior, and from Saratoga to the Mississippi, includes numerous fossil shells, which, in some of the strata, are almost exclusively of a single species of Terebratula; other species of this genus, equally numerous in individuals, form other strata, mixed with Encrinites, Alcyonites, Caryophillites, Favosites, Gyrogonites, &c. In another stratum is found the Alcyonite, Trilobite, a Terebratula with flat valves, a Favosite, &c.; and a large discoidal shell, which more particularly forms the subject of this paper, and which at first sight resembles an Ammonite, or Nautilus; and was noticed as such by Mr. Maclure in his geological observations, page 27. We first observed an impression of it in the compact limestone which forms a portion of the bank of Lake Erie, near Eighteenmile creek, mixed with Caryophillœa, and subsequently at Basin Harbour, on Lake Champlain, where several more perfect individuals occurred; several good specimens were lately sent by a gentleman of Kentucky, to the Philosophical Society, one of them, an impression, 10 or 12 inches in diameter, and another exhibiting a perfect vertical section.

Mr. Samuel Hazard, a member of the Academy, presented to our cabinet a collection of fossils collected by himself, in Kentucky, amongst which were some specimens of this shell. Sometime afterwards Mr. Clifford, of Kentucky, presented me with fine specimens which he brought from Tennessee river.

A careful examination of all these individuals, in their several states of preservation, presented to me the common characters of a discoidal form, flat spire, very large umbilicus, and entire cavity; the last trait distinguishes them from Nautilus. The genera to which it makes the nearest approach are Solarium and Delphinus of Lamarck, but it is separable from them by the characters by which I indicate this new genus.

Genus *MACLURITE.

Generic Character.

Shell discoidal, much depressed, unilocular; *spire* not elevated, flat; *umbilicus* very large, with a groove formed by the projection of the preceding whorls, not crenulated.

Species.

1. M. *magna*. *Shell* obtusely carinated on the exterior upper edge; *whorls* rapidly increasing in size; *aperture* on the left, irregularly oval, horizontally depressed above; *lips* not reflected.

My Cabinet, Cabinet of the Academy, and of the Philosophical Society.

Plate 13, fig. 1. Upper surface of the M. magna exhibiting *a*, the remains of the shell, *b*, the cast, *c*, thickness of the shell.

Fig. 2. A vertical section—*a*, outer lip, *b*, umbilicus.

Fig. 3. Under part of the shell.

2. M. *bicarinata*. *Whorls* acutely carinated on the middle above, and obsoletely carinated beneath; *aperture* on the right.

Cabinet of the Academy.

Parkinson's Organic Remains, vol. 3, page 76, pl. 5, fig. 1 and 3.

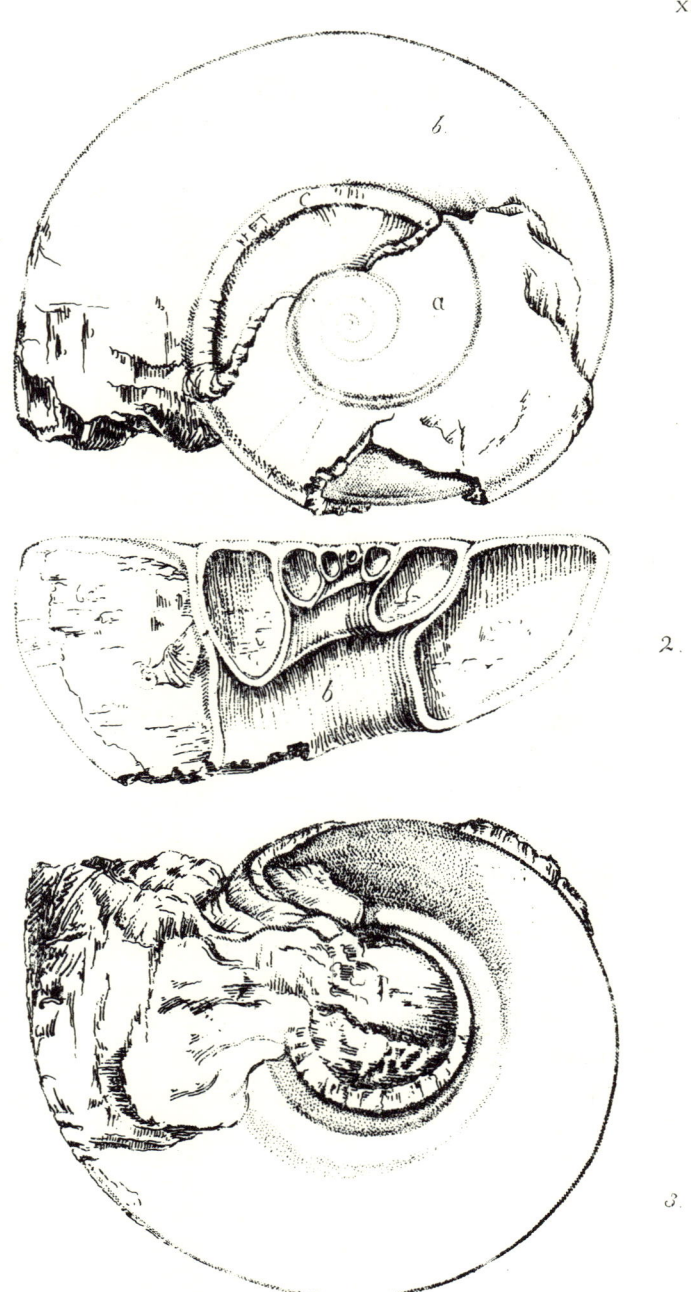

Part III
MEDICAL GEOLOGY

Editor's Comments on Papers 8 and 9

8   **MITCHILL**
        Outlines of Medical Geography: Being an Inquiry How Far Calcareous Soils and Strata Counteract the Septic Exhalations Which Occasion Distempers of a Febrile or Pestilential Type

9   **ANONYMOUS**
        Reviews of Six Treatises on Mineral Springs of the United States

Many of America's first natural scientists were educated as physicians and applied their studies of natural productions to the improvement of health. Medical geography—or the study of the relationships between bedrock, soils, drainage features, and the prevalence of disease—was of special importance to the expanding American nation. The New York doctor Samuel Latham Mitchill (1764-1831) was among the first to emphasize the need for care in selecting building sites and materials. Mitchill was an active scientist, notable for the diversity as well as the number of his publications. His writings include essays on botany, anatomy, medicine, zoology, and the earth sciences. In addition to studies on the geological structure of New York and Long Island, he contributed articles on iron ores, mineral springs, earthquakes, fossils, mineral localities, and medical geography.

In "Outlines of medical geography" (Paper 8), Mitchill discusses the salubrious qualities of regions underlain by calcareous as opposed to argillaceous sediments. In Mitchill's day some diseases of the "febrile or pestilential type" were known to result from exposure to the "septic exhalations" of sewage. Because these septic exhalations were thought to be acidic, Mitchill believed calcareous rocks and soils would neutralize the cause of disease. Application of these ideas to British geology led him to speculate that towns situated on limestone or chalk formations would have a lower rate of illness than those underlain by alluvial or organic soils. Carrying these ideas one step further, Mitchill (1801) extolled the "excellence of calcareous material for building and paving cities, particularly as respects their power to overcome the exciting cause of fevers." It is not evident, however, that his opinions affected the architecture or urban planning of the expanding American nation.

Mineral springs and waters attracted considerably more attention from medical researchers before 1820. Health centers in Virginia and at Saratoga and Ballston, New York, were well known by 1800, but the origin and even the validity of their healing virtues remained a matter of debate. The growing number of conflicting reports on the composition and capabilities of mineral waters prompted the editors of the *New England Journal of Medicine and Surgery* to review six studies on New York springs (Paper 9).

Chemical analysis of these mineral waters was an important avenue of research. If the healing properties of the springs were due to a chemical property, then healing waters produced in the laboratory might be made available to all without the need for difficult overland journeys. As early as 1793 Valentine Seaman published "A method of making artificial mineral water, resembling that of Saratoga" based on his analytical experiments, and many others attempted similar ventures. Analysis of the solid and gaseous ingredients was no simple task, however, as demonstrated by the *New England Journal's* summary of inconsistent results by several noted chemists.

Even as these mineral waters were being analyzed, the salutary influences of the springs were seriously questioned by some physicians. The health of many visitors improved, it was acknowledged, but it was not resolved whether these benefits were due to a property of the water or, rather, to "change of air, exercise, temperance, and the influence of expectation." Whatever the reasons, North American mineral springs remained popular health resorts throughout the nineteenth century.

REFERENCE

Mitchill, S. L. 1801. Excellence of Calcareous Materials for Building and Paving Cities, Particularly as Respects Their Power to Overcome the Exciting Cause of Fevers. *Med. Repos.* 4:91–93.

BIBLIOGRAPHY

Baltzell, John, 1802. *An Essay on the Mineral Properties of the Sweet Springs of Virginia, and Conjectures Respecting the Process of Their Production by Nature.* Baltimore, Werner and Hanna, 30p.

Cooper, T. 1813. On Mineral Waters, and Watering Places: Particularly the Carlisle and York Springs in Pennsylvania, with a Method of Making Artificial Mineral Water. *Emporium of Arts and Sci.* n.s. 1:474–481.

Drake, D. 1816. Medical Topography (of Cincinnati). *Eclectic Repos. and Anal. Rev.* 6:137–149.

Hosack, D. 1810. Answer to Dr. Seaman's "Examination" of a Review of His Dissertation on the Saratoga and Ballston Waters. *N.Y. Med. Philos. Jour. and Rev.* **2**:145-175.

Hosack, D. 1810. Observations on the Use of Ballston Mineral Waters in Various Diseases. *Am. Med. Philos. Regist.* **1**:42-47.

Rouelle, J. 1792. *A Complete Treatise on the Mineral Waters of Virginia Containing a Description of Their Situation, Their Natural History, Their Analysis, Contents, and Their Use in Medicine.* Philadelphia: Charles Cist, xix, 68p.

Seaman, Valentine. 1808. An Examination of the Account of an Analysis of the Balltown Waters. *Med. Repos.* **11**:253-256.

Tenney, S. 1793. An Account of a Number of Medicinal Springs at Saratoga, in the State of Newyork (sic.). *Am. Acad. Arts. Sci. Mem.* **2**:43-61.

OUTLINES OF MEDICAL GEOGRAPHY:

Being an Inquiry how far Calcareous Soils and Strata counteract the septic Exhalations which occasion Distempers of a febrile or pestilential Type. In a Letter from Dr. MITCHILL, *F. R. S. E. Member of the Legislative Assembly, and Secretary of the Agricultural Society of the State of New-York, Professor of Chemistry in Columbia College, Fellow of the Societies of Philadelphia, Boston, &c. &c. to* JAMES HAWORTH, *M. D. and Radcliffian Travelling Physician from the University of Oxford, dated November* 4, 1797.

IF there is any solidity in the observation made in my letter to Dr. Beddoes of September 15, 1797, concerning the comparative mildness or rarity of pestilential and febrile distempers in countries underlaid with extensive strata of superficial *lime-stone*, as happens in some parts of the United States, then England ought to exhibit something of the same kind, in the counties where *chalk* is very prevalent. I now set myself down to recollect what I remarked about calcareous earth in travelling from Dover to London, and during a walk I once took from London to Oxford and its environs, Woodstock, Blenheim, and back again.

The chalk cliffs in the neighbourhood of Dover, are talked of by all persons who navigate the Channel up and down, or who barely pass from any of the opposite ports of France to this part of England. It has not been so generally noticed that the land from Folkstone, in Kent, or thereabout, along by South-Foreland, and almost to Deal, together with good part of the Isle of Thanet, including both Ramsgate and Margate, has a chalky bottom. Indeed, with very little interruption, the calcareous material prevails to a very considerable extent on both sides of the road from Dover to Canterbury, and thence to Rochester. In short, I may say it extends westward quite to the county of Surry, and passes into it by a tract almost as broad as the distance from Deptford to Westerham. This is the most healthy part of the country, as may appear by comparing it with the unwholesomeness of Oxney Isle, Romney Marsh, and the Isle of Sheppey, where not chalk, but siliceous sand, flint, clay and loam constitute the principal part of the soil. (See the map prefixed to Mr. BOYS' General View of the Agriculture of the County of Kent. London. 1796.)

The chalk continues through part of Surry, Berkshire, much of Oxfordshire, quite into Gloucestershire, and prevails extensively in Buckinghamshire, Middlesex, and somewhat in Essex. Oxfordshire, though situated so far inland, is famous in many places for its petrifactions and incrustations of calcareous earth. Through this cretaceous land run the *Isis*, the *Charwall*, the *Windrush*, the *Evenlade*, the *Thames*, and, according to Dr. PLOT's enumeration, *seventy* other streams of inferior rank, whose collected waters pass through the enumerated chalky tracts to London, and the greater part of the distance thence to the sea.

The latter of these rivers, and some others which fall into it below, are prettily noticed by POPE in his poem of Windsor Forest.—

"———————— in ancient times we read,
Old Father Thames advanc'd his reverend head:
His tresses dropp'd with dews, and o'er the stream
His shining horns diffus'd a golden gleam:
Grav'd on his urn appear'd the moon, that guides
His swelling waters and alternate tides;
The figur'd streams in waves of silver roll'd,
And on their banks *Augusta* rose in gold;
Around his throne the sea-born brothers stood,
Who swell with tributary urns his flood;
First the famed authors of his ancient name,
The winding *Isis* and the fruitful *Thame*;
The *Kennet* swift, for silver eels renown'd;
The *Loddon* slow, with verdant alders crown'd;
Cole, whose dark streams his flowery islands lave;
And *chalkey Wey*, that rolls a milky wave;
The blue transparent *Vandalis* appears,
And gulphy *Lee* his sedgy tresses rears;
The sullen *Mole* that hides his diving flood,
And silent *Darent*, stain'd with Danish blood."

The inferences to be drawn from these geographical facts are two. 1st. That an extensive body of chalk, or calcareous earth, of very considerable breadth, tinctures, with its peculiar qualities, the soil of England from the German Ocean westward, to the hills which separate the waters of the Severn from those of the Thames in Gloucestershire. 2dly. That the greater part or all of the streams falling into the Thames, run through a country charged with calcareous earth, and, consequently, must be considerably impregnated with that material.

With regard to the first point, if the *septic gas*, or *effluvium*, is, as experiments lead us to believe, of an acid nature, in that case it ought not to be very abundant or long prevalent as a cause of epidemic, malignant, and pestilential distempers, where the land is composed, in a good proportion, of calcareous earth; or, if in any place, or at any time, it should happen that febrile disorders of greater or less malignity or inveteracy break out, this ought to happen in consequence of too much acid present for the chalk or lime to neutralize. On looking over the first authority that comes in my way, MARTIN's Natural History of England (edition published in 1759), I find Surrey denominated generally " a pleasant country;" Middlesex commended for its " exceedingly healthful air;" Berkshire called " one of the most pleasant counties in England;" Buckinghamshire said to possess an air generally good, especially on the Chiltern Hills; and though the vale is dirty, "'not so unhealthy as many other low lands in England, THE SOIL BEING MARLE OR CHALK;" and the eastern part of Gloucestershire, where the Evenlade and Windrush rivers arise, though less fertile than the western part of the county along the Severn, and more exposed to wind and cold, " makes amends by its healthfulness."

The healthfulness of Oxfordshire is almost proverbial, and the judicious choice of King ALFRED has long been commended for pitching upon so wholesome a spot as that where the beautiful city of Oxford stands, as a seat for the muses. This has been considered as very strikingly manifested by the two following circumstances. 1st. An observation of long standing, that although the small-pox is as frequent there as any where else, the effects of it are seldom fatal; and, 2d. That when the pestilence, in 1665, was spread in a manner over the whole kingdom, though the court, both houses of parliament, and the terms, were held in the city of Oxford, yet the plague, notwithstanding, was never there at all. (1 Martin's, &c. p. 368.)

On the other hand; where, along the fenny hundreds of Essex, the joint operation of the current of the Thames, and the tides of the ocean, have brought together great bodies of sand and clay, that had been diffused through the water, and been deposited by it, together with all such remains of animal and vegetable matter as constitute the deep mould of those tracts, the country abounds thereabout with septic exhalations, which are very injurious to the health of the inhabitants, where the neutralizing power of calcareous earth or chalk is wanting: and, as far as I can judge, a deficiency of the same material enters deeply into the explanation of the unhealthiness of the fens of Lincolnshire, and the adjoining hundreds of Fleg and Mershland, in Norfolk.

It must be observed, however, although chalky and calcareous soils are thus destructive of that kind of air or vapour which produces agues, fevers, &c. that it does not follow, by the rule of contraries, that all sandy, loamy, gravelly, and clayey soils must be unhealthy. This may, however, be remarked, that in the United States of America, the most sickly parts are the tracts along the Atlantic, where the land consists, in pretty much the whole range, from New-York to Florida, of silicious sand and clay, variously mingled, and tempered with more or less of mould, without any considerable admixture of calcareous earth; and this, when it occurs, consisting principally of concretions of marine shells, dug up in some parts of Virginia and the other southern States. As to our cities, the site of New-York is a sandy loam, or gravel, except that part where the plague has usually prevailed hitherto, which is built upon salt meadow, miry swamp, and rotten trash. Philadelphia stands partly upon a sandy, and partly upon a stiff loamy or brick-earth, very impermeable to water, and destitute of a sufficient proportion of calcareous earth to keep down and attach pestilential fluids. Though, of both these cities, it may be remarked, with truth, that though they were built upon a bottom of *lime*, or *chalk itself*, there are local causes enough to engender the worst forms of distempers, as happened at the famous assizes in Oxford, where the filth, accumulated around the wretched criminals in prison, generated pestilential matter enough to poison a considerable number of the court and attendants.

As far as I can comprehend the subject, the general conclusion is this: Countries abounding with calcareous earth are mostly free from the ravages of wide-spreading epidemics, by reason of the power that material possesses of neutralizing the acid of putrefaction; though, from particular local causes, pestilence may be manufactured in places thus favourably circumstanced: while those places which are most remarkable for the prevalence of malignant and pestilential diseases of that kind, consist of a soil through which chalk and lime are scantily scattered. England affords abundant proof of the former, the United States of the latter, observation.

To proceed now to the second point: The numerous streams of water whose united body forms the Thames, run principally, it was said, through a tract of country plentifully furnished with chalk. The form in which this calcareous earth is most prevalent is that of carbonate, or in combination with fixed air, and capable, in many places, of being burned to quicklime: though, doubtless, where the septic (nitric) acid exists, it unites with the earthy basis into calcareous nitre. There is, unquestionably, somewhat of a muriate of lime, especially below London bridge, and the whole distance thence through the brackish and salt-water to the ocean. And perhaps there may be, in some places, a combination of the sulphuric acid with the calcareous matter into gypsum. The small quantity of gypsum which exists there, and if much did exist, the very trifling proportion of it which water is capable of dissolving, leave very little to be said concerning it here. The muriate of lime is there probably in too small proportion to be matter of any moment any where above the flow of the salt-water; and though, as CAMDEN says, the river swells as high as Richmond, which is sixty miles inland, (Britannia. 187.) with the tide, yet the salt reaches not a great distance from Gravesend.

The septite of lime is soluble in water; but as it is also a nutritive ingredient in manures, and easy of resolution into its constituent parts by plants, it is presumable that a large proportion of what is formed of that compound undergoes decomposition on the soil where it is produced. The carbonate of lime, then, or common mild chalk, is the material upon which the streams of water, in the case before us, must principally act.

Water may work upon this mild calcareous earth in two ways: 1. By acting as a menstruum, or jointly with some substance which is a menstruum, and *chemically* dissolving it; and, 2. By *mechanical* attrition, wearing it away, reducing it to fine particles, and while it is diffused and floating through the liquid, carrying it down in its proper form to be mingled and deposited with the various matter of intervale spots, alluvial shores, and secondary islands.

It is hard to determine what quantity of calcareous earth is brought down in these two modes, from the adjacent counties into the bed of the Thames, and hurried along with his stream towards its disemboguement. When all circumstances of showers, rains, freshets, and the unceasing atteration of the streams along their channels, are taken into the account, the quantity would seem to be very considerable. This chalky substance, whether dissolved in the water, or diffused through it, cannot fail to modify and influence it remarkably; in an especial manner contributing to fertilize the low lands in the neighbourhood of the river which it visits or overflows.

Thus the septic and other acids with which the Thames water was charged, are, in a great measure, or perhaps quite, neutralized by the calcareous earth, before it reaches London. The quality of the water there will be, in a good degree, determined then by the materials furnished by that city. The acid of putrefaction and other acids running into it from the sewers, will, under the existing regulations, be neutralized by the refuse alkaline matter of house-keeping, and other consumptive processes; and the potash and soda thrown away in soap, &c. seem to more than counterbalance their opponents, and give a preponderating influence to the alkalies. Hence evidently proceeds the boasted softness of the Thames water for the washing of linen; for enabling the London dyers to strike their bright and lasting colours; for keeping so well at sea as it does; and, probably, enabling it to extract, in a very complete manner, the substance of malt to form their excellent porter. Is not all this confirmed by the fact known to navigators, that the Thames water, after long keeping, deposits, within, a stony sediment on the bottoms of the casks?

So much for the waters of the Thames, and the chalky regions through which they run. The facts are very striking and interesting, but they are not the only ones of the kind.

When I was in France, I once passed up the river Loire, partly by land, and partly by water; but, at Angers, turned off to the left on my rout to Paris. I therefore did not travel through that fine tract of country watered by the Mayenne, the Indre, and Upper Loire, which abounds in the soft calcareous stone, that by bare exposure to the open air, becomes highly charged with septic acid. Nor did I visit Saumur, the center of a district of calcareous earth, in which the acid of putrefaction is so abundant as to require nothing but an addition of pot-ash to turn the calcareous nitre to salt-petre. If the philosophers of France would inquire into the circumstances with a due degree of attention, I am persuaded they would find that the lime-stone, which, by combining with this acid, gives to the soil much of its fertility, prevents, at the same time, the rise of its pernicious vapours into the atmosphere. (FOURCROY's Speech to the Council of Ancients, Med. Repos. No. I.—3 Diction. Univ. de la France, Artic. SAUMUR, p. 64, where the salt-petre refined there is called the best in France.)

From the manner in which Mr. HOUSEMAN expresses himself in the Journal of his late tour through England, there is strong reason to believe that the salubrious air, for which Matlock in Derbyshire is so famous, has a connection with, and is partly dependant upon, the lime-stone-rock, abounding in the valley through which the river Derwent there runs (Monthly Mag. for March, 1797. p. 203.)

But that fertile tract of land, THE CARSE OF GOWRIE, in the county of Perth in Scotland, which is reckoned to possess a climate more mild and favourable to vegetation than any part of that kingdom, affords *direct* evidence of the healthiness consequent upon using lime as a manure. The soil consists chiefly of rich clay, loam, and sharp gravel; and the inhabitants, until the year 1735, used to be subject to the ague. Then one or two of the principal proprietors undertook, by draining, summer-fallowing, and sowing grass-seeds, to improve their estates. Accident led them to a discovery of the efficacy of lime on that soil, from observing the powerful effects of *some old lime rubbish of decayed buildings*, when spread on the corner of a field. The liming their lands then gradually came into use, and has since been generally adopted; the consequence of which is, the AGUE HAS LONG

AGO DISAPPEARED. Here seems to have been a beautiful experiment made upon about ninety-six square miles of country, where the septic steams that formerly gave the people agues, are now attracted by the lime and turned to calcareous nitre, while increased productiveness of the land, and greater wholesomeness of the air, continue to be the happy consequences. (Donaldson's General View, &c. p. 12. and seq.) Some judgment may hence be formed concerning the power of art in changing the face of nature. What a grand reflection, that an inconsiderable quantity of powdered lime strewed over the land, should thus coerce the matter of pestilence, and controul the operations of the atmosphere!

I shall bring this letter toward a close by inserting some facts of the same kind, afforded by Sicily. *Calcareous earth* abounds in the neighbourhood of Palermo, in the forms of *lime-stone*, (2. Swinburne's Travels, p. 199.) *stalagmites*, (p. 201.) and *breccia*, (p. 203). On the road thence to Girgenti, it exists in the form of *marine concretions, talk, gypsum,* (p. 227.) *chalky stones,* (p. 243.) and *rocks,* (p. 256). In the neighbourhood of this place, the ancient Agrigentum, *marine exuviæ,* constitute good part of the lower strata of the hills, (p. 258.) and many of the buildings are constructed of a conglutination of sea-sand and *shells,* (p. 268). On the road from Alicata toward Syracuse, some of the cliffs bordering on the Mediterranean Sea, are composed of a *greenish marle,* full of sulphur and *solid rocks of gypsum,* (p. 284). And on the road to Messina, the traveller meets with lofty broken mountains, composed of marble and various sorts of calcareous stones, (p. 358.) and with high calcareous *cliffs of red and white marble,* (p. 361). To this basis of calcareous earth, which is now covered over, to great extent, by lava and other volcanic productions, that remarkable island owed formerly, and still owes a great share of its wholesomeness and fertility. The manner in which this was effected, seems visible on the *Leontine Plains* and their vicinity. The town of Lentini, though situated on a spot so exceedingly productive of vegetables, is, however, unhealthy during summer and autumn, on account of its nearness to the lake of Biveri, and a great space of country filled with lakes and ponds. The hills which inclose it on the east side, afford great quantities of saltpetre; and people are constantly employed in scraping it off the walls. Do not the vapours of the low grounds which infect the atmosphere there, attach themselves to an alkaline basis as soon as they meet with it, and form the nitre? Is not the formation of the nitre, in such circumstances, a tolerable indication that a prevalent ingredient in the air of that sickly region, is the septic (nitric) acid vapour exhaled from the place of its production, the marshes? And is not the effluvium which exhales from the head of the port of Syracuse formed in like manner, by chemical union of the septon (azote) and oxygene of substances undergoing putrefaction there; which, volatilized by the heat of summer, vitiates the air and endangers the lives of the inhabitants? And was it not a miasma of the same nature and composition, which caused that malignant fever, or plague, which destroyed the Carthagenian army encamped in the fens, when they came to rescue this city from the Romans?

But why do I take the tour of Europe to establish the fact of the salubrity of countries consisting of calcareous earth, or underlaid by it? Nearer home, the island of Bermudas and its neighbouring isles, furnish abundant proof of the same thing. The whole cluster consists of calcareous matter, apparently formed of the fragments of marine shells compacted into a porous and friable sort of white stone. This neutralizes the acid of their rain water, and all other acids, so effectually, that a generally pestilential state of the atmosphere is unknown there; and the Somers' islands have become famous for the wholesomeness of their climate, and the longevity of their inhabitants. Yet even there, damaged provisions, excessive use of salt meat and fish, wheaten flour, and meal of maize, both soured by keeping, sometimes excite considerable febrile sickness, which, however, is remarked never to spread by infection; the septic acid evolved from meat and meal being checked by the lime as soon as it is wafted abroad, but being still capable of working its accustomed mischief in many of the places and substances where it is immediately produced.

One paragraph more and I shall conclude. Read the hypothesis of Linnæus (1 Amænitat. Academic.) on the cause of intermittent fevers, and you will find a collection of facts to prove their connection with argillaceous earth, or clayey soil. Of this he was so well satisfied, that he concluded that attenuated particles of clay taken into the body with food and drink, entered the blood, stuck in the extreme branches of the arteries, and brought on, as a true proximate cause, the symptoms of the disease. (Hypothesis nova, § v.) The sensible inquirer will find, in his fourth section, an enumeration of all the parts of Sweden famous for intermittents and strata of argillaceous soil; and the authority of Mr. Sandel, quoted as an eye-witness of the same coincidence of clayey bottoms and intermittent fevers in Pennsylvania. The facts I take to be indubitable. Linnæus has reasoned upon the subject, by considering argillaceous earth alone. I have viewed it in contrast with calcareous earth, that, by embracing a wider range of facts, the operation of the latter, in tempering the former, may be the better comprehended. Whether my theory is better founded than that of the Professor of Upsal, the experienced and candid will judge.

I thank you for your copy of old MAYOW. That this man, who was a London physician, and a fellow of All-Souls-College, his brilliant discoveries, and his book which contains the account of them, should all be forgotten, in his own country as well as abroad, within less than one hundred years, so effectually that FRANKLIN could never have read his explanation of water spouts, nor SCHEELE his detection of dephlogisticated air, nor GIRTANNER his manner of accounting for muscular action, nor any body else, what he has left on respiration and on the condition of the unborn fœtus, and unhatched chick, are among the most singular occurrences in the literary history of the 18th century.

Wishing you a prosperous voyage, and a happy meeting with your friends, and with ALMA MATER, I finish my letter, by assuring you, &c.

SAMUEL L. MITCHILL.

REVIEWS OF SIX TREATISES ON MINERAL SPRINGS OF THE UNITED STATES

Anonymous

1. *A Dissertation on the Mineral Waters of Saratoga, including an Account of the Waters of Ballston.* By Valentine Seaman, M. D. *New York*, pp. 131. 1809.
2. *Analysis of Ballston Water, communicated by Dr. Hosack to Dr.* Miller. *New York Med. Repos. Hex. 2d, vol.* v. 1808.
3. *Observations and Experiments on several Mineral Waters, in the State of New York.* By Mr. John Griscom. *Bruce's Min. Jour.* vol. 1, p. 156.
4. *Chemical Examination of the Water of the Congress Spring, Saratoga.* By J. F. Dana, M. D. *New Eng. Jour. Med.* vol. vi. p. 19.
5. *An Experimental Inquiry into the Chemical Properties and Medicinal Qualities of the principal Mineral Waters of Ballston and Saratoga, in the State of New York, &c.* By William Meade, M. D. *Mem. of the Am. Phil. Society of Philadelphia, &c. &c.* 8vo. pp. 167. *Philadelphia,* 1817.

6. *Analysis of the Mineral Waters of Ballston and Saratoga, &c. &c.* By Dr. John H. Steel, 18mo. Albany, 1817.

BALLSTON and Saratoga are now become such fashionable places of resort during the summer season, either for amusement or the restoration of health, that the public at large feel some curiosity to know the circumstances, which are thus sufficiently powerful to draw together such numbers of visitors from all parts of the United States. In order to gratify, so far as is in our power, so laudable a wish, we have collected all the publications which have come to our knowledge, and from these we shall endeavour to present to the reader, an account of the country in which these springs are found, of the nature of the substances contained in their waters, and of the uses to which they have been applied.

The village of Ballston is situated in the county of Saratoga, state of New York, about two hundred miles north of the city, and twelve miles west of the Hudson river; it is in a deep valley, surrounded by a range of undulating hills, which form a kind of amphitheatre, and is intersected by a branch of the river Kaydarosoras. The soil is naturally poor and sandy, bearing oak, pine, and hemlock; but in the vicinity of the village it has been rendered more productive by the use of plaster of Paris. The ground is principally composed, says Dr. Meade, of two or three species of rocks, of secondary formation, but these are so covered with immense beds of sand, that it is difficult to ascertain this formation, and it can only be done by an attentive examination of the rivulets, which in some places have laid bare the strata. Large blocks of quartz, and rolled masses of primitive rocks, are distributed over the surface of the ground in and about the village, which circumstance perhaps gave rise to the opinion of Drs. Seaman and Mr. Griscom, that there are many rocks at Saratoga, &c. approaching in hardness to porphyry; but we believe, with Dr. Meade, that these are to be considered as accidental, and by no means to determine in any degree the geological character of the place. Mr. Griscom affirms that neither calcareous nor magnesian stones are to be found in the neighbourhood; but Dr. Meade, who appears to have studied the structure with much attention, has discovered, that where the stratum of soil or sand is washed away, an entire range of flœtz, or horizontal rocks is laid open to view, consisting of a species of schist which is nearly black, which stains the fingers and effervesces slightly with acids. It appears to contain both carbon and carbonate of lime, and to approach in character the aluminous schale. It breaks into columnar or prismatic masses, and its fractured surfaces present the impressions of a species of grass. Within three miles of Ballston, solid masses of calcareous rock may in some places be observed, particularly near the site of a mill, where the flœtz formation is beautifully illustrated. This limestone is nearly of a black colour; its fracture is slaty; it abounds with shells of various forms, some of which are so very apparent in their structure and form as not to be mistaken; they principally consist of bivalves, madrepori, terebratulites, corrolites, and echinites; so exceedingly similar, says Dr. Meade, to fossils in my possession from Mendip, in England, that it is difficult to distinguish the specimens from each other. When rubbed, this stone emits the odour of the *pierre puante*. The rocks of Saratoga present the same general characters, but the shells are not so abundant, and they are penetrated by veins or seams of flint or chert. These observations of Dr. Meade appear unequivocally to prove the geological character of the vicinity of the springs. According to Dr. Steel, sulphuret of iron, plumbago, and small specimens of fluor spar are found in the neighbourhood.

These villages owe their celebrity to the mineral waters, which are poured forth most copiously from many parts of their surface. It appears from Dr. Steel's work, that they had been known, from time immemorial, to the aborigines of the country, whose attention, according to their tradition, was first attracted towards them by the great quantity of game by which they were frequented; and this notice was afterwards kept alive by the salutary influence of the waters upon some of their diseases. In the year 1767, Sir William Johnson, who suffered from the gout, was persuaded, through the recommendation of the Indians, to visit the spot and drink the waters from the fountain. He passed through the wilderness, remained some time in the neighbourhood of the spring, and returned renovated by health. Some time after Sir William's return, a man by the name of Norton, influenced by the growing celebrity of the spring, obtained permission, from the owners of the soil, to erect buildings and clear the land in the vicinity. Under this permission a small hut was built, and a clearing made; but the fear which the hostile Indians inspired, during the revolutionary war, which commenced about this time, induced him to abandon his improvements. After the contest was decided, Norton returned to his old abode, and farther settlements were made in 1784 and 1785, by others, by whom the three springs, now known by the names of Flat

Rock, President, and Red Spring, were probably discovered. The discovery of similar springs at Ballston, about the year 1787, checked the growth of the settlement at Saratoga, and nothing of consequence was done until the year 1803, when the valuable qualities of the Congress Spring having been duly appreciated, Mr. Putnam was induced to erect a large house for the convenience of those who were allured thither to drink the waters. Since that time, says Dr. Steel, the waters of Saratoga have gradually gained a celebrity that appears as firmly as it is justly established. Numerous houses for the reception of company have been erected, and the influx of persons the present season, from all quarters of the Union, has, we understand, much exceeded the amount of any former year.

At Ballston there are three springs, which have gained notoriety. The first, or the Public Well, so called from having been reserved for the benevolent purpose of serving the public by Sir W. Johnson, in the original grant of the land to private individuals, is situated nearly in the centre of the village. This spring, says Dr. Meade, issues from a bed of stiff blue clay and gravel, which lies near a stratum of schist or shale, nearly on a level with the brook or rivulet, which runs through the town, and the course of which has been diverted from the springs by a dyke or mound. A circular vessel of wood forms the well, in which the water rises to within four feet of the surface, and is dipped out in glass tumblers fixed to rods, a method which is certainly inconvenient and awkward, and which might be easily remedied. The inhabitants have surrounded the well with a marble pavement, supporting an iron balustrade.

The second spring rises about two hundred yards from the public well, and is the property of an individual. It is situated lower in the valley, not many feet above the level of the rivulet; the ground around it is not high, and it is covered with a building, which contains conveniencies for warm bathing.

The third which, from the presence of sulphuretted hydrogen, has acquired the name of the Sulphur Spring, rises within twenty feet of the one last described.

By digging in the neighbourhood of the two last wells in almost every direction, springs will be found, says Dr. Meade, exhibiting the same qualities in a slight degree, but by no means so strong as the waters of Lowe's spring, which, with the public well already mentioned, are the principal ones that are generally and indiscriminately used. "The situation of the country round Saratoga differs but little from that of Ballston, except that the hills are not so high, and the valley is more extensive. It lies low, and the soil is principally sand, or gravel, covered with peat. In this valley a number of springs are to be found, showing more or less the same sensible qualities. To these wells different names are given, such as the Congress Spring, the Flat Rock, the Hamilton, the President, the Columbian, the Round Rock, &c. They are but a short distance from each other; and, as the same appearances present themselves in every part of this valley, many more may certainly be found if it was examined."

" The water in each of those springs arises from a bed of sand, intermixed with stiff blue clay, and overlaying the calcareous and schistose rocks.

" The source of these springs does not appear to lie deep, as they are all found within five or six feet of the surface."

" They are all confined or enclosed in circular or square wooden vessels, not more than five or six feet deep."

" One of these springs, called the Round Rock, has something so peculiar in appearance as to attract particular notice. It is situated at the foot of a calcareous rock, low in the valley, and is covered by a cone or pyramid near six feet high; this cone is hollow, and has an opening at top about nine inches wide, from which the water can be seen in a state of agitation, as if boiling, from the extrication of gas, which rises to the surface. An opening at the bottom of this cone, about four or five inches wide, on a level with the surface of the ground, gives exit at present to the water." This cone is supposed by Dr. Meade, and we presume correctly, to have been formed by the gradual deposition of the carbonate of lime, which this water is known to contain. pp. 19 and 20.

The waters of all these springs, except the one which is impregnated with sulphuretted hydrogen, coincide in their characters, differing from each other only in degree. When viewed at the wells, the surface appears in continual agitation, and a vast number of small air bubbles are seen to rise and break. Small animals, immersed in the air of the wells, are soon suffocated. When first taken from the springs, they are perfectly transparent and colourless; they emit a number of air bubbles, sparkle when poured from one vessel to another, and their taste is first pungent, afterwards saline, and finally ferruginous; on being allowed to remain at rest for a few minutes, especially in warm weather, the gas which they contain separates, and covers the whole internal surface of the glass with minute bubbles; they then soon lose their transparency, a pellicle, slightly iridescent, appears upon the surface; a cloudiness, or opacity, ensues, and finally there takes place a precipitate of a light brown powder, which adheres firmly to the glass. After this process the water resumes its transparency, but its taste is saline and disagreeable. Such are the sensible qualities of the Ballston and Saratoga waters. We shall now present the reader with the results of a number of analyses, executed by different chemists, in order to determine not only the nature but the proportions of foreign matter, to the presence of which they owe their medicinal qualities.

In looking over these analyses we have found that experiments have been made upon different quantities. Some chemists have given us the solid and gazeous contents of a pint, some of a quart, and others of a gallon. In comparing them, therefore, with each other, the reader, in order to determine the relative proportions, will have to reduce them to a common measure, and, as this is somewhat of an irksome business, we have undertaken to relieve him from the task, and give the amount in *one hundred cubic inches* of water.

BALLSTON.

1. *Public Well.*

100 cubic inches of water yielded of
Muriate of soda	72·72 grains.
Muriate of magnesia	3·03
Muriate of lime	5·62
Carbonate of lime	16·01
Carbonate of magnesia	20·34
Oxide of iron	1·73
	118·75
Carbonic acid	105·62 cubic inches.
Azote	4·59
	110·21

Spec. grav. of the water 1008
Temperature 52° Fah.

MEADE.

Muriate of soda	68·83 grs.
Carbonate of soda	3·89
Carbonate of lime	32·68
Carbonate of magnesia	1·08
Carbonate of iron	3·03
	109·51

Carbonic acid	90·90 cub. inch.
Temp. 50 Fah.	

STEEL.

Muriate of soda	65·30
Carbonate of lime	46·36
Muriate of magnesia	26·34
Muriate of lime	10·53
Carbonate of iron	8·44
	156·97
Carbonic acid	300 cub. inch.

ANONYMOUS.*

2. Low's Well.

Muriate of soda	74·46 grs.
Carbonate of lime	23·37
Carbonate of magnesia	12·98
Muriate of lime	7·35
Muriate of magnesia	4·76
Oxide of iron	1·73
	124·65
Carbonic acid	104·76 cubic inches.
Azote	4·59
	109·35

Spec. grav. 1008.
Temp. 52° Fah.

MEADE.

---

* The author of this analysis is not known; it was made at Paris, as it is said, by an eminent chemist, and the results were published in the New York Med. Repos. for 1808. The quantity of water employed amounted to 25 ounces, which, if of Troy weight, is equivalent to 47·450 English cubic inches, and he obtained 31 grains of common salt. Hence it is easy to find the proportion in the common measure we have adopted; for as $47·450 : 31 :: 100 : x = 65·30$, and in the same way may be ascertained the relative proportions of the other substances. Of this analysis we shall say something hereafter.

Muriate of soda	- -	61·47 grs.
Carbonate of soda	- -	4·32
Carbonate of lime	- -	27·92
Carbonate of magnesia	- -	0·64
Carbonate of iron	- -	2·59
		96·94
Carbonic acid	- -	95·23 cub. inch.

Temperature 52° Fah.

STEEL.

SARATOGA.

1. *Congress Spring*.

Muriate of soda	- -	178·35 grs.
Muriate of lime	- -	5·62
Muriate of magnesia	- -	8·22
Carbonate of lime	- -	47·61
Carbonate of magnesia	- -	29·43
Oxide of iron	- -	0·86
		270·09
Carbonic acid	- -	114·28 cub. inch.
Azote	- -	3·46
		117·74

MEADE.

Muriate of soda	- -	204·11 grs.
Carbonate of lime	- -	77·26
Carbonate of soda	- -	7·14
Carbonate of magnesia	- -	1·45
Carbonate of iron	- -	2·67
		292·63
Carbonic acid	- -	148·48 cub. inch.

STEEL.

Muriate of soda	184·74 grs.
Carbonate of soda	6·06
Carbonate of lime	62·33
Carbonate of magnesia	12·64
Silex	2·59
	268·36

DANA.

Sulphate of soda.
Muriate of soda.
Carbonate of soda.
Carbonate of magnesia.
Carbonate of iron.
Silex.

CUTBUSH.

2. *Flat or Round Rock Spring.*

Muriate of soda	70·99 grs.
Carbonate of lime	26·83
Carbonate of magnesia	17·74
Muriate of magnesia	6·92
Oxide of iron	0·43
	122·91
Carbonic acid	114·28 cub. inch.
Azote	1·73
	116·01

MEADE.

Muriate of soda	84·32 grs.
Carbonate of soda	4·84
Carbonate of lime	46·32
Carbonate of magnesia	0·64
Carbonate of iron	3·24
	139·36
Carbonic acid	109·31 cub. inch.

STEEL.

Carbonate of soda	- -	9·00 grs.
Muriate of soda	- -	59·91
Carbonate of lime	- -	65·80
Carbonate of iron	- -	2·94
		137·65
Carbonic acid		69·29 cub. inch.

SEAMAN.*

The waters of other springs have also been analyzed, particularly by Drs. Meade and Steel, but the above, we presume, will furnish sufficient examples to the curious reader, upon which to institute a comparison, not only as regards the nature, but likewise the proportions of the substances which constitute the active ingredients of these mineral waters. Now, then, let us look a little into these analyses, and endeavour to make out the points in which these gentlemen coincide or disagree.

A single glance at the results of these analyses, is sufficient to shew that they differ exceedingly from each other, not only in relation to the salts, which are the products of evaporation, but to the relative proportions in which they exist. Of the muriates, Dr. Meade finds those of soda, lime, and magnesia, while Dr. Steel denies that the waters contain any other than common salt; on the contrary, the latter discovers carbonate of soda, which could not be detected by the former. The French chemist obtains two muriates and two carbonates. So much for the waters of Ballston. The analyses of those of Saratoga are equally discordant: Drs. Meade, Steel, Dana, and Seaman, find carbonate of Soda; Dr. Meade alone discovers the muriates of lime and magnesia, and Mr. Cutbush sulphate of soda. They all agree that the carbonates of lime, magnesia, and iron, and the muriate of soda, invariably constitute a part of their solid contents.

If we now turn our attention to the numbers which represent the proportions in which these saline bodies are present, we shall find them equally at variance; and, to state this fact more simply and clearly, the decimals shall be rejected, except

---

* Dr. Seaman's analysis was made on the solid contents of ten pounds of the water, and we have estimated the above proportions upon the supposition that one pound (or 12 ounces Troy) is equivalent to 28·875 cubic inches.

when the quantity of a substance is very small. Let us take, then, as examples, the muriate of soda, the carbonates of lime, magnesia, and iron, and carbonic acid.

BALLSTON.

1. *Public Well.*

	M.Soda.	Carb.Lime.	Carb.Mag.	Oxid.Iron.	Carb.Acid.
Meade	72	16	20	1·73	105
Steel	68	32	1·08	3·03	90
French	65	46		8·44	300

2. *Low's Well.*

	M.Soda.	Carb.Lime.	Carb.Mag.	Oxid.Iron.	Carb.Acid.
Meade	74	23	12	1·73	104
Steel	61	27	0·61	2·59	95

SARATOGA.

1. *Congress Well.*

	M.Soda.	Carb.Lime.	Carb.Mag.	Oxid.Iron.	Carb.Acid.
Meade	178	47	29	0·86	114
Steel	204	77	1·45	2·67	148
Dana	184	62	12		

2. *Round Rock.*

	M.Soda.	Carb.Lime.	Carb.Mag.	Oxid.Iron.	Carb.Acid.
Meade	71	26	17	0·43	114
Steel	84	46	0·64	3·24	109
Seaman	59	65		2·94	69

It is difficult to account for the great difference observed in the proportions above stated. Very considerable errors must exist somewhere; for the amount of the whole saline mass obtained by evaporation, does not vary in any great degree, and it is therefore improbable that the strength of the waters should, at different times, fluctuate to such an extent as to afford Dr. Steel 77 grains of carbonate of lime, and Dr.

Meade only 47 grains; or that the latter should find 29 of carbonate of magnesia, where but one grain and a half were obtained by the former. We can more readily believe that the two carbonates might have been confounded, or that the acicular crystals of sulphate of lime, formed by the analytical method which was pursued, might, perhaps, have been mistaken for sulphate of magnesia.

It is strange, also, considering the general uniformity in the modes of analysis, that one should detect the muriates of lime and magnesia, and find no trace of carbonate of soda, while three others state the carbonate as one of the ingredients, but are totally silent respecting the muriates. In order to reconcile these opposing authorities, it will be necessary to look into the analyses, and endeavour, if possible, to discover the source of these discordant results. First, then, of *carbonate of soda*. It was discovered by Dr. Dana, by pouring upon the dry solid contents of the Congress Spring water, two ounces of pure water; the solution, after filtration, was evaporated to dryness; upon the mass was poured distilled vinegar, which caused an effervescence; the solution was evaporated to dryness, mixed with alcohol of the specific gravity of 815, and, after standing a sufficient time, was filtered. The insoluble part was muriate of soda. The alcoholic solution was then evaporated to dryness, and the mass, by exposure to heat, was converted into a carbonate, which, when dissolved in water, and purified by nitrate of silver from a small portion of muriate, proved by subsequent examination to be carbonate of soda. Dr. Steel, on adding infusion of red cabbage to the solution which was formed by pouring water upon the mass obtained by evaporating a certain volume of the mineral water, states, " that it gave the whole a beautiful green colour; and, having previously ascertained the number of grains of carbonate of soda required to neutralize a given number of drops of dilute muriatic acid, he added this acid, drop by drop, to the water, until the original colour of the infusion was restored, and then, by subsequent evaporation, he obtained nothing but entire crystals of muriate of soda. This experiment, although rather coarse compared with the method usually practised to prove the presence and the proportion of carbonate of soda, was sufficient to shew its existence. Dr. Meade obviously detected the presence of an alkali, which must have been the carbonate of soda; for, in the analysis of the water of Low's well, he found that paper, stained with turmeric, had its colour changed to an orange brown; but, prepossessed by an opinion, which he derived from Bergman and Kirwan, that fixed alkalies and earthy salts are incompatible with each other, he concluded that the alkali which produced this change in turmeric, existed only in minute quantity, and neglected to ascertain either its nature or proportion; (p. 35). There is every reason to believe, however, that analysis does not always afford the ingredients which pre-existed in the water; that salts, which have been supposed incompatible as sulphate or carbonate of soda, with muriate of lime or magnesia, may exist as such, at a certain degree of dilution, and that, as the solution is concentrated, a mutual exchange of principles may take place, and new salts be formed. If, therefore, the muriates of lime and magnesia were actually dissolved in the water, they might re-act upon the carbonate of soda during the evaporation, and the resulting products would be muriate of soda, and carbonate of lime or magnesia, or both. This mode of viewing the subject might satisfy us respecting the reason why Dr. Meade failed in obtaining carbonate of soda; but then, directly in the face of this argument, we have the assertion of Drs. Seaman, Dana, and Steel, and of Mr. Cutbush, that this salt is actually obtainable, in notable quantities too, from the solid contents of the Saratoga waters, and there appears no reason to doubt of their correctness. How to reconcile these opposite results we know not, and shall therefore leave the mystery to be cleared up by those who are more immediately interested in the business. In the second place Dr. Meade invariably found, in his alcoholic solutions, muriates of lime and of magnesia. Dr. Seaman and Dr. Dana saw no indications of these salts, and as to Dr. Steel, he roundly asserts that alcohol, purified by muriate of lime, could not be made, by him, to take up the smallest quantity of the saline property from the residuum of any of the waters, p. 93. This discordance is very extraordinary. One would suppose, that, in an experiment of this kind, which consists only in adding alcohol to a saline mixture, filtrating and subsequently evaporating, no difference in opinion could possibly exist, whether a residuum was or was not obtained. Yet here one chemist not only finds that two salts are dissolved, but, by certain processes, fixes with precision their relative proportions; while another is equally clear, that all this must be an illusion, and that no salts soluble in alcohol can be procured from the solid contents of these waters. The analysis of the French chemist favours the conclusions of Dr. Meade; but we know not the authority upon which, in this instance, we may depend, and the gross and palpable error which has been made in the estimate of the volume of carbonic acid, and the weight of the carbonate of iron, would lead us to doubt the correctness of the whole examination. The weight of evidence against the existence of the muriates in the residue of these waters is, when numbers are considered, as four to one; but as we cannot at present reconcile these opposing authorities, we shall take the old adage as an apology, and be silent. It is possible that carbonate of soda, and the muriates, even perhaps that of iron, may exist together in the waters of the wells, and that, during their concentration, they mutually decompose each other, the salt of soda being present in larger proportion than is necessary to produce this effect, and the excess consequently constituting a part of the residuum. As for the *sulphate* of soda, discovered by Mr. Cutbush, we think there must have been some mistake; for all those who have examined the waters, although they disagree in almost every thing else, are unanimous in the opinion that no indications of sulphuric acid are to be perceived. We shall now say something of the quantity of oxide or carbonate of iron, and of carbonic acid, procured by these gentlemen. On looking at the tables, it will be seen, that, from the waters of the public well, Ballston, Dr. Meade obtained 1. 73 gr.; from Low's the same quantity; from the Congress spring, Saratoga 0·86 of a grain, and, from the water of Flat or Round Rock, 0·43 of a grain. From the same waters and the same volumes, Dr. Steel succeeded in separating 3·03; 2·59; 2·67; 3·24 grains of oxide of iron; whence it will appear, that where Dr. Meade found the smallest, Dr. Steel discovered the largest quantity; and, conversely, that, where the latter separated the smallest, the former procured the largest quantity. Dr. Meade's method of obtaining the iron was, to expose the solid contents of the water (e. g. of Low's well) to the action of air and of moisture for three weeks, until the iron was "*reduced*" to the highest state of oxidation, and then to separate the other ingredients by the action of acetic acid. The metal is thus converted into the peroxide, and the author quotes Mr. Chenevix as saying, that it is then in the fourth or highest state of oxidizement, and weighs as 189 to 100 of metallic iron. We were not a little surprised at finding that Dr. Meade took the oxide, thus formed, as the same with that which existed in the water, and that he should have copied such a mistake as the one which he gives on the authority of Mr. Chenevix. He exposed the protoxide of iron, for three weeks, to the air, in order that, by converting it into the peroxide, it might be rendered insoluble in acetic acid. During this change, it absorbed one proportion of oxygen, and the peroxide, thus formed, was weighed and recorded in his table, as the quantity found in the water. Dr. Meade ought to have known that there are but two oxides of iron,* the black and the red, the former being considered as the protoxide, and the latter as the peroxide; and that, in each case, the metal being represented by 100, the oxygen, in the first, must be considered as 30, and in the last as 45, and not 89, as he has stated. The real quantity of oxide of iron, dissolved in the water, would amount to about 0·90 of a grain. The same correction must be made in the weight of oxide stated in the other analyses. Dr. Steel appears to have estimated the quantity of iron by the weight of the precipitate produced by prussiate of lime; but the details of the process are not given. From the great difficulty of forming and of preserving the simple prussiate, we are inclined to suppose that the triple prussiate was employed, in which case, unless an allowance were made for the iron existing in it, the amount of this metal in the water must have been overrated. This perhaps may account for the difference in the weight of this ingredient observed in his and Dr. Meade's statements.

With respect to the volume of carbonic acid contained in the waters, a very considerable difference will also be found to exist between the numbers of Drs. Meade and Steel. The former, in his examination of four of these liquids, states them as follows: 105, 104, 114, 114 cubic inches; while the corresponding analyses of the latter give 90, 95, 148, 109, in 100 cubic inches of the water. The method practised by Dr. Meade to separate this volatile product, was to fill a vessel of a certain capacity with the water, and to invert over its orifice a decanter graduated into cubic inches, and filled with water at the temperature of 120°; the vessel was then placed in a sand bath, the contents were made to boil, and the ebullition was continued until the gas ceased to be extricated. After allowing the water and the air in the receiver to cool to 60°, the number of cubic inches of water, which had been displaced, was noted, and the purity of the gas was tried by lime water. Dr. Steel caught the gas in a bladder, and then passed it into a graduated vessel, filled with water, at the temperature of 70°. In three analyses out of four, he obtained less of carbonic acid than Dr. Meade; and here the difference, which is not very

---

* M. Guy Lussac asserts that he has discovered a third oxide, containing oxygen in proportion intermediate between those of the black and red; but the fact has been disputed.

considerable, may be accounted for by the probable absorption of a portion of gas, both by the bladder (for we presume it was moist) and by the water, through which it was passed. But from the water of the Congress spring, Dr. Steel obtained no less than thirty-four cubic inches more of carbonic acid than Dr. Meade. Our mode, therefore, of explaining this discordancy must here fail, and we are under the necessity of again referring for some explanation, on this head, to the gentlemen who are most interested in the result.

Dr. Steel affirms, that the elastic fluid, procured by his process, was pure carbonic acid, and that no other gaseous product could be detected. By Dr. Meade, on the contrary, it is stated, that a notable proportion of azote was found mixed with the carbonic acid; and he has accordingly, in his analysis, uniformly recorded this gas as one of the aerial ingredients of these mineral waters. Although azote has been found in similar waters in Europe, and we think it not very improbable that it may be present in those of Ballston and Saratoga, the mode adopted by Dr. Meade to obtain it, was not so satisfactory as to convince us of the truth of his assertion. The gas was caught over water heated to 120°; after cooling, it was repeatedly passed through fresh lime water, and, after being in contact with the liquid until the absorption ceased, there remained a small proportion of gaseous fluid, which was inferred to be azote, because it extinguished flame. Now we believe that the atmospheric air, contained in water, is not all evolved by heating it to 120°; to effect this separation it requires boiling. Supposing, then, that a portion of common air remained, it would be displaced by the carbonic acid, and would reassume its elastic form, and, when mixed with one-seventh of its volume of carbonic acid, it would be rendered incapable of supporting combustion. This, however, we offer merely as a conjecture. It would have been easy to have ascertained the presence of atmospheric air, by the mutual action of nitrous gas, and the oxygen which it contains; an experiment which is performed over water, and, consequently, does not require the aid of a mercurial apparatus, with which it appears Dr. Meade was not supplied.

But it is time to turn our attention to a subject which will probably be considered by a majority of our readers of much more consequence than that with which we have hitherto been occupied; and that is, the medicinal qualities of these waters. We shall, therefore, close this view of the analysis with the remark, that the discordance exhibited throws a shade of obscurity over the real ingredients of these waters, and that it would puzzle the physician, who sends his patients to the springs to decide, whether *he* prescribes muriates or carbonates, or *they* are to drink any thing more than a solution of common salt, chalk, magnesia, a little iron, and a great deal of carbonic acid.

It seems, however, to be agreed by physicians in Europe, who have attended to the medicinal effects of mineral waters, that their operation is more salutary than that of any artificial combination, possessed of the same chemical properties. We shall not, therefore, deny to the mineral waters of Ballston and Saratoga the credit of many wonderful cures of various diseases, marked by totally different and even opposite characters; nor do we much care to ascertain, whether these cures are wrought by change of air, exercise, temperance, the influence of expectation, or by some mysterious virtue in these natural combinations. The following are Dr. Steel's remarks on the effects of the waters in the diseases, for which they have been most celebrated:

" The waters are so universally used, and their effects so seldom injurious, particularly to persons in health, that almost every one who has drank of them assumes the right to direct their use to others, and even empiricks, without any knowledge of their composition, and little or none of their effects, contrive to dispose of their *directions* to valetudinarians, to no other purpose than to injure the reputation of the waters, and destroy the prospects of the diseased.

" Nothing can be more absurd than the idea that governs many who visit the *springs* for the restoration of their health, that they are to recover in proportion to the *quantity* they drink; for although persons in health may, and frequently do, swallow down enormous quantities of the water with impunity, it by no means follows, that those whose stomachs are enfeebled by disease, can take the same quantity with the same effect. Stomachs of this description, most frequently, reject the too copious draught, and save the system from the evil consequences that would otherwise inevitably follow; but when it happens to be retained, the result is indeed distressing; the pulse becomes quick and feeble, the extremities cold, the bowels swollen and painful, and the whole train of nervous affections alarmingly increased; and, should the unfortunate sufferer survive the effects of his imprudence, it is only to a renewal of his worst apprehensions from a loss of confidence in what he most probably considered a last resort.

" Among the great variety of invalids who resort to the springs, none, perhaps, receive more essential and effectual benefit from their use, than *the Bilious and Dyspeptic.* In the first, if the attack be recent, and unattended with any serious organic affection, it is most usually removed, in the course of a few days, by a free use of the Congress water; but in those cases where the functions of the stomach and bowels have become impaired, from the long continuance of the disease, attended with anasarcous swellings of the extremities, &c.; although the waters of this fountain may be resorted to with nearly the same assurance of obtaining relief, nevertheless, more caution is indispensably necessary in its administration; for, should a great quantity of the water be drank, without having the proper effect by the bowels and kidneys, it is never beneficial; but, on the contrary, frequently increases the most alarming symptoms of the disease. In this case I have been in the habit of recommending the conjunction of some mild cathartic medicine; and, for this purpose, two or three grains of *calomel* have been given over night, followed in the morning with three or four tumblers of the water, with the happiest effect; a few doses of this description, usually places the bowels in a situation to be more easily wrought upon by the water alone, and the patient becomes convinced of its efficacy in his disease, from a few days proper application.

" In Dyspepsia, it is usual to begin a course of the waters with the Congress. This should be taken in the morning before breakfast, four or five tumblers full are commonly sufficient to produce a pretty copious discharge from the bowels, and, in weak irritable habits, half the quantity, or a single tumbler full, in some cases, is amply sufficient to answer the purpose; but in those cases where the bowels are attended with an habitual constipation, the quantity of water required to move them is apt to prove too cold to the stomach, and, by producing cold chills and nausea, frequently defeats the general intention of its application; this may be prevented by taking some suitable laxative over night, and a much less quantity of water in the morning will answer the wishes of the patient, without subjecting him to any inconvenience. But the water of the Congress is not, alone, to be depended on for the removal of this disease; when the stomach and bowels have been sufficiently cleansed by the pleasant and innocent purgative properties of this water, recourse must be had to the operation of the more powerful *chalybeates;* these are to be found in the waters of the *Flat Rock,* the *Columbian,* in *Ellis's Spring,* and at the *Spa.*

" The quantity of waters, from either of these fountains, to be used daily, depends, in a great measure, on the state of the disease and the disposition of the stomach; it is therefore necessary to commence their use in small quantities at a time, in distant and regular intervals, gradually increasing the quantity and frequency of the draught, as may be most agreeable to the stomach, and least injurious to the feelings. In this way, the quantity may be increased to from one to two quarts a day, and it is questionable whether a much larger quantity may be drank with any additional advantage.

"Conjoined with the internal use of the water, bathing should not be forgotten; its exhilarating effect upon the surface, contributes much to the restoration of the vigour and health of the stomach. The cold shower bath should always be preferred where the energy of the system is sufficient to overcome the effects of the cold, and produce the sensation of warmth over the surface of the body, immediately after its application. Where this sensation is not produced, the cold bath should be dispensed with, and the tepid, or warm bath, substituted in its stead, together with general friction, with a flesh brush or coarse flannel, over the whole surface.

"The stimulating effects of these waters, arising from their saline and gaseous properties, give them a decided preference over any other, as a bath; and those who are labouring under a deficient or irregular action of the cutaneous vessels, arising either from a sympathetic affection with a diseased stomach, or from an original affection of the vessels themselves, will find it to their advantage to persevere in its use under this form.

"The idea of bathing before sunrise, or early in the morning, is entirely erroneous. Before bathing, the system should always receive the invigorating effect of moderate exercise and a nutricious repast. The hour of ten or eleven in the forenoon is, therefore, the most suitable time for its application.

"In calculous and nephritic complaints the waters have long been celebrated for their efficacy; and numerous well attested instances of their good effects can be produced, where the disease was not only mitigated, but effectually cured. In these cases the subjects of them voided large quantities of sand and small gravel; and, for some years past, have felt no symptoms of the return of the complaint.

"The fountain that would seem to promise most in these diseases, are the Hamilton, and Taylor's Washington, as they contain the greatest quantity of the *aerated alkali*. But the waters have been usually drank indiscriminately for this purpose, without reference to any particular fountain; it is, therefore, probable, that the fixed air and lime add to the *lithontriptic* properties of these waters.

"They should be drank in such quantities as to keep the bowels loose, and repeated sufficiently often to keep up an increased secretion by the kidneys.

"In chronic rheumatism, the virtues of the waters were celebrated by the aborigines, and later observations confirm the justice of their faith. The Congress water has the most celebrity in this disease. It should be drank in the morning, in sufficient quantities to move the bowels two or three times, followed, through the day, by moderate draughts of some of the other fountains; and, in most instances, the *shower bath* will add much to the efficacy of the water. Following this course, for a length of time, gradually relaxes the rigidity of the muscles, adds strength, and facility of motion, to the diseased joints, and restores ease and vigour to the whole system.

"*Scrofula* is another disease for which those who are afflicted with it frequently become applicants to the waters, and experience has sanctioned the belief of their utility in this afflictive complaint. The chalybeate waters are those from which we are to look for the greatest benefit; they must be commenced in small doses, and the quantity gradually increased as the stomach will bear them, and their use continued at least through the summer months. There are but few of this description that have not received advantage; and numerous instances might be adduced, where the less seriously affected have perfectly recovered, in consequence of a proper course of bathing and drinking.

"In dropsy, arising from viceral obstructions of long continuance, the waters are manifestly injurious, as they invariably increase the swelling and add to the sufferings of the patient; but, in recent cases, where the affection arises simply from a deficient action in the absorbent vessels, the water has a singular effect in removing it; it should be drank in the morning freely, so as to produce a copious discharge from the bowels, and, through the day, taken in such quantities as to keep up a pretty constant discharge of urine. The bloating is relieved immediately, and a subsequent course of chalybeates, will finally establish the permanency of the cure.

"In *Paralysis* the waters have usually been singularly serviceable; the purgative properties of the Congress render it most applicable to this disease, and its good effects are much increased by the use of the bath.

"In chlorosis, and a variety of other complaints peculiar to the female sex, the waters maintain a *high* and *deserved* reputation. In these cases the bowels should be kept loose by the use of the more purgative waters, and the stronger chalybeates should be persevered in for a length of time; their good effects will be accelerated by frequent bathing and moderate exercise."

Dr. Meade not only presents us with a very copious account of the medicinal properties of the spring waters, but also occasionally steps out of his way to give his views of diseases and modes of cure. In treating of dyspepsia, he has afforded us some novel and very entertaining remarks on a complaint which has of late years appeared in Boston, which he has thus described:

" In Boston, in particular, of late years, a disorder has prevailed, to which the faculty have given the name of dyspepsia; to doubt the propriety of which, would seem to be an instance of some presumption. As far as my observations have gone, a complaint somewhat similar is there a very frequent disease. It principally makes its attacks at a very early period of life, reducing the patient to the utmost state of emaciation and debility: many of the symptoms are such as are usually observed in dyspepsia; they are continued for a length of time, and are aggravated in the winter months. Children of the age of ten or twelve years are not exempted from it, and some of the finest young men in the country are attacked with it from the age of fifteen to twenty-one. It soon reduces the patient to a state of the utmost debility and emaciation, such as is usual in the last stage of phthisis pulmonalis; the countenance is altered, the cheeks are prominent, the eyes look hollow and languid, the hair often falls off, the nails are of a livid colour, and the pulse becomes so low, that I have, in one instance, been able to count only forty-five strokes in a minute. The persons most liable to this complaint are remarkable for a particular formation; a long neck, prominent shoulders, narrow chest, clear skin, and thick upper lip, with other marks of a scrophulous diathesis. They have no cough, nor can any suspicion be entertained of an affection of the lungs during the whole of the disease. If I am correct in those appearances, have we not much ground for suspicion that there is a scrophulous disposition in the system, and that the whole of those symptoms arise from an obstruction in some of the lymphatic glands in the neighbourhood of the stomach, or in those of the mesentery, similar to incipient tubercles of the lungs, and rendering them incapable of conveying chyle or nourishment to the system? thus arises that emaciation which constitutes a species of marasmus, and is so similar to that which occurs in the latter stage of consumption."

Now we have no doubt that Dr. Meade may actually have seen such a train of symptoms as he describes, at least in two or three individuals. The *extreme debility* and *emaciation*, the *altered countenance*, the *prominent cheeks (cheek bones)*, the *hollow* and *languid eyes*, the *falling hair*, and the *livid nails*, have actually appeared, and Dr. Meade may have witnessed them, in at least *two* and possibly *three* instances. But after casting our eyes about us, to ascertain whether a disorder, with these symptoms, to which the faculty have given the name of dyspepsia, has really *prevailed* in Boston of late years, we must confess that none such has been discovered. The disease which the Boston physicians have ventured to call dyspepsia, does not make its attacks *at a very early period of life*, but generally in subjects between twenty-five and forty-five years of age. It is not *aggravated in the winter months*, but is usually mitigated in these months, and troublesome in hot weather. This complaint is often relieved by travelling, and by change of climate, but not by change to a *warm climate*. The persons who have had this disorder *have not been remarkable for any particular formation;* nor can we trace a common resemblance between them in any one point. This dyspepsia has appeared in young men of five and twenty, who have been careless of their regimen; in robust gentlemen of forty-five, who have habitually lived too well; in married ladies, whose constitutions have been impaired by inattention to exercise, and too great an exertion to nourish their infants. People with thick upper lips, narrow chests, and clear skins, do not appear to have been more troubled with the dyspepsia than those with thin lips, dark skins, and black hair. Hence it would seem probable that the disease in question is not of a scrophulous nature; a probability which is strengthened by the facts, that, in the moist climates of Great Britain and Holland, where scrophula is the most common of diseases, we do not find enrolled in the long list of dyspepsiæ *symptomaticæ* and *idiopathicæ* of Cullen and Sauvages, any disease under the name of Dyspepsia Scrophulosa; while, in the dry air of Boston, where this new kind of dyspepsia is said to exist, it appears that scrophulous disorders are comparatively rare. We, therefore, venture to conclude, with deference to the authority and experience of the author,

1. That this disease is not of a scrophulous nature, since it does not occur in scrophulous climates nor constitutions.

2. That, since it arises from the same causes as dyspepsia; exists in the same kind of subjects; exhibits the same symp-

toms, and is cured by the same means—it is neither more nor less than a genuine dyspepsia.

If it be asked, whether indigestion be actually more prevalent in Boston than other American towns, where good living is equally common, we should make our reply with due circumspection. We should allow, that a few severe and remarkable cases have existed, and excited much attention both here and at the Ballston springs; but might add, that we have noticed ladies and gentlemen, of other cities than Boston, who were travelling about in search of health, under the severe affliction of *bilious* complaints, which, to our eyes, exhibited a considerable resemblance to the Boston dyspepsia. It would look like presumption to say, that, in such cases, the mischief is more likely to proceed from a weak stomach, than a corrupt liver; but, at least, it will be generally admitted by impartial people, that of all our organs, there is none more commonly ill used and over-strained than the gastric viscus. Possibly we might be disposed to go a little farther, and allow, that the stomach, in the inhabitants of Boston, is endowed with a more obstinate and reluctant character than that of other people; whence, instead of exhibiting its revenge for abuse, in a slow and gentle manner, on our ligaments and membranes in the form of gout, or on the liver in the shape of *bilious* complaints, it enters at once into a contest with the articles of food that are presented to it, and either repels them without ceremony, or converts them into biting acids, and horrible explosive gases, which produce the formidable train of symptoms described by Cullen—" Anorexia, nausea, inflatio, ructus, ruminatio, cardialgia, gastrodynia," &c. and sometimes even the more formidable phenomena, so eloquently represented by Dr. Meade.

These treatises are meritorious for the industry and science they exhibit, and the judicious advice to the public, respecting the use of the mineral waters; but it is to be regretted, that they are all deficient in a geological map of the country in which the springs are situated.

A Treatise on Verminous Diseases, preceded by the natural history of Intestinal Worms, and their origin in the Human body. By VALERIAN LEWIS BRERA, *Professor of Clinical Medicine in the University of Pavia, ornamented with five plates. Translated from the Italian, with notes, by* Messrs. J. BARTOLI, M. D. &c. *and* CALVET. Paris, 1814.

Part IV
MINERALOGY

Editor's Comments
on Papers 10, 11, and 12

10 ANONYMOUS
Excerpt from *Mineralogy*

11 BRUCE
Description, and Chemical Examination of an Ore of Zinc, from New Jersey

12 SILLIMAN
Review of An Elementary Treatise on Mineralogy and Geology Being an Introduction to the Study of These Sciences, and Designed for the Use of Pupils; for Persons Attending Lectures on These Subjects; and as a Companion for Travellers in the United States of America—Illustrated by Six Plates, by Parker Cleaveland

To understand the history of mineralogy, one must understand the development of mineral classification. Perhaps more than in any other branch of the earth sciences, progress in the study of minerals depended on an unambiguous method for identifying and arranging objects for study. Mineral classification presented a tremendous challenge to natural historians of the eighteenth and early nineteenth centuries. A given mineral may appear dissimilar in two different localities, whereas different minerals may seem identical to the untrained eye. Structurally related species, such as members of the pyroxene group, may be chemically different, whereas such chemically similar minerals as orthoclase feldspar and muscovite mica may be unrelated in terms of structure. Chemical analysis, furthermore, was exceedingly difficult before 1820, and the science of crystallography was unknown. It is not surprising, therefore, that Americans made little progress in mineralogy before 1800.*

European researchers in the late eighteenth century proposed several classification schemes based on the physical, chemical, and morphologi-

---

*John Greene and John Burke (1978) have chronicled the history of mineralogy in early America, and the reader is directed to their work for a more comprehensive account.

cal characteristics of minerals. In Germany, Abraham Gottlob Werner systematized the grouping of minerals by "external" or physical properties. The French chemist Lavoisier proposed a primarily chemical organization of minerals at nearly the same time that his fellow countryman René Just Haüy introduced a classification of minerals based on crystal morphology.

The near-simultaneous emergence of three distinct bases for mineral classification resulted in no one preferred system. On the contrary, many mineralogists felt obliged to publish their own combination of physical, chemical, and morphological traits for mineral identification (Burke 1969). Werner, Haüy, Kirwan, Jameson, Brochant, Brongniart, Karsten, and Lucas are among the authorities whose systems are cited by Parker Cleaveland in his 1816 American mineralogy text. Mineralogical nomenclature was confusing at best in the early nineteenth century.

The lack of a standardized system of mineral nomenclature and identification was not the only difficulty confronting American mineralogists. Mineralogy is not a science that can be learned from treatises and textbooks alone; laboratory and field examinations under the guidance of a trained observer are essential for gaining expertise in mineral identification. American progress in mineralogy was retarded more by the lack of teachers than for a want of a good textbook. Of the years before 1800, Benjamin Silliman (Paper 12) remarked:

> It was a matter of extreme difficulty to obtain, *among ourselves*, even the *names* of the most common stones and minerals; and one might inquire earnestly, and long, before he could find any one to identify even quartz, feldspar, or hornblende.

John Greene (1969) emphasized the importance of cooperative efforts in the development of the science:

> The early history of American mineralogy is a story of the activities of individual collectors and investigators located in towns and cities from Georgia to Maine; of their relationships with roving European scientists; of their efforts to organize for the exchange of ideas, the pooling of books and specimens, and the publication of researches; and, finally, of growing cooperation in the common task of exhibiting the mineral products of America to the republic of science and letters.

Examination of American publications in mineralogy without consideration of these facets of the mineral sciences may yield an incomplete view of the development of this discipline.

The first major mineralogical work to be printed in North America was an English translation of the system by the Swedish chemist Cronstedt, published with additional commentary on physical and chemical identification procedures. Cronstedt's system of classification, which predated those of Werner, Lavoisier, and Haüy, was based on such qualitative chemical properties of minerals as reaction to water, oil, and heat. The chemical basis of mineral classification was favored by the anonymous English editor, who claimed:

> It is but of late, since the principles of chemistry were well understood, that mineralogy has been advanced to any degree of perfection. The best way of studying mineralogy, therefore, is by applying chemistry to it; and not contenting ourselves merely with inspecting the outsides of bodies.

This preference for a chemical classification of minerals long persisted in North America and culminated in James Dwight Dana's great *System of Mineralogy.*

Cronstedt's system of mineralogy must have been widely available in America by 1800. The Philadelphia edition of Cronstedt's treatise was first published as part of the 18-volume *Encyclopaedia, or Dictionary of Arts, Sciences, and Miscellaneous Literature,* a publication that was purchased by libraries and families throughout the continent. In addition, a book-form version of the text was republished in at least two Philadelphia editions as the *Compendius System of Mineralogy* (Paper 10). In spite of this availability, the 1794 text appears to have had little impact on the progress of American mineralogy. Only occasional reference is found to Cronstedt in later North American publications (see, for example, Paper 18B p. 276). Perhaps the lack of trained chemists and mineralogists prevented the more widespread use of this forgotten work.

Significant progress in establishing North American mineralogy as a science was made after the introduction of chemistry courses in several medical schools. Physicians' interest in minerals was not purely academic, for these natural chemicals provided some of the raw materials for medications. The first American chemistry courses were taught in Philadelphia by Dr. Benjamin Rush, and several of his students, notably James Woodhouse, John Redman Coxe, and Adam Seybert, applied their expertise to minerals. The efforts toward an unambiguous chemical identification of minerals were greatly aided by another Philadelphian, Robert Hare, who in 1801 introduced his "compound" or oxyhydrogen blowpipe (Hare 1802). Previous blowpipes produced insufficient temperatures to fuse many silicates and other oxygen-based minerals, but Hare's device was effective in determining the melting properties of these refractory compounds.

By 1800 quantitative mineral analyses were appearing regularly in American periodicals such as the *Medical Repository* and the *Transactions of the American Philosophical Society.* The rapid progress of American mineral analysis was further stimulated by the publication of an American edition of Frederich Accum's *Practical Essay on the*

Analysis of Minerals. With active investigators and collectors in most Eastern cities, the American mineralogical community was well prepared to support a periodical devoted exclusively to the earth's natural productions. Archibald Bruce (1777-1818), a New York doctor and professor of materia medica and mineralogy at Queens College, commenced the *American Mineralogical Journal* in 1810 and served as its editor until 1814 when the first volume was completed. Although the journal had a brief life because of financial difficulties, its pages reflect the dynamic and growing science of mineralogy in North America. Whereas fewer than a dozen qualified mineralogists resided in the United States and British North America before 1800, more than 24 researchers contributed to this publication.

The state of American mineralogy is well illustrated by Bruce's own contributions to his journal. His "Description, and chemical examination of an ore of zinc, from New-Jersey" (Paper 11), a brief but detailed note, is of special interest as one of the earliest descriptions of a new mineral species by an American. The "red oxide of zinc," now known as zincite, was one of the principal ores at the famed Franklin, New Jersey, mines, and specimens of this spectacular mineral are to be found in most natural history museums.

Mineral collectors made substantial contributions to the development of mineralogy in America, both by their labors in discovering new and unusual specimens and by their generosity in making these specimens available for study. Early American periodicals contain notices of many private collections or "mineral cabinets," including those of Archibald Bruce, Benjamin De Witt, David Hosack, William Maclure, Benjamin Perkins, Adam Seybert, and John Webster. Perhaps the most notable American collection of its day, however, was that of Colonel George Gibbs (1776-1833), who "amassed more geological and mineralogical specimens than any other individual of the age" (Eaton 1820). Gibbs was a wealthy gentleman who brought his valuable collection of European specimens to America in 1805. His efforts in mineralogy were not purely acquisitive, for he was a frequent contributor to Bruce's *American Mineralogical Journal* and was honored as the first vice president of the American Geological Society (Paper 30). The Gibbs cabinet of 10,000 minerals eventually became the core of the Yale University collection and, in conjunction with the teaching talents of Benjamin Silliman, was an important factor in Yale's dominance in mineralogy throughout much of the nineteenth century.

The culminating achievement of American mineralogy before 1820 was the publication of Parker Cleaveland's *Elementary Treatise on Mineralogy and Geology* in 1816. The significance of this textbook goes far beyond the fact that it was the first such publication by an American, for Cleaveland developed his own classification scheme based on chemical tests consistent with the state of the art of American mineral analysis. This classification, coupled with a compilation of American mineral localities, resulted in a textbook uniquely suited to encourage and advance mineralogy in the United States and Canada.

Cleaveland's success was quickly acknowledged in more than a dozen favorable reviews in American and European periodicals (Papers 12 and 29). Benjamin Silliman demonstrated his high regard for Cleaveland's *Treatise* by including his laudatory review as the second article in volume 1 of the new *American Journal of Science,* now the oldest North American periodical in continuous publication. Silliman emphasized Cleaveland's contribution by outlining the difficult beginnings of American mineralogy and the lack of a satisfactory system of mineral classification prior to the *Treatise.* The prediction that Cleaveland's work would "greatly promote the knowledge of mineralogy" was borne out, for the *Elementary Treatise* in its two editions became the standard textbook in American schools for almost two decades.

REFERENCES

Burke, John G. 1969. Mineral Classification in the Early Nineteenth Century. In *Toward a History of Geology,* edited by C. J. Schneer. Cambridge, Mass.: MIT Press, pp. 62-77.

Eaton, A. 1820. *Index to the Geology of the Northern States,* 2d ed. Troy, N.Y.: W. S. Parker and Albany, Websters and Skinners, xi, (1), (13)-286p., 2 plates.

Greene, J. C. 1969. The Development of Mineralogy in Philadelphia, 1780-1820. *Am. Philos. Soc. Proc.* **113**:283-295.

Greene, J. C., and J. G. Burke. 1978. The Science of Minerals in the Age of Jefferson. *Am. Philos. Soc. Trans.* **68**, pt. 4:113p.

Hare, R. 1802. *Memoir on the Supply and Application of the Blow-pipe.* Philadelphia: Chemical Soc. of Philadelphia.

BIBLIOGRAPHY

Anonymous 1818. (Review of Parker Cleaveland's *Elementary Treatise on Mineralogy and Geology*). *Am. Mon. Mag. and Crit. Rev.* **1**:183-187.

Anonymous 1817. Review of Parker Cleaveland's *Elementary Treatise. Analectic Mag.* **9**:301-314.

Bruce, A. 1814. Description of Some of the Combinations of Titanium Occurring Within the United States. *Am. Mineralog. Jour.* **1**:233-243, plate.

Channing, W. 1817. Review of Parker Cleaveland's *Elementary Treatise. North Am. Rev.* **5**:409-429.

Cleaveland, P. 1816. *An Elementary Treatise on Mineralogy and Geology.* Boston: Cummings and Hilliard, xvi, 668p., 6 plates.

Cutbush, J. 1812. *Hydrostatics, or a Treatise on Specific Gravity.* Philadelphia: for the author, 49p.

De Witt, B. 1804. An Account of Some of the Mineral Productions of the State of New York. *Am. Acad. Arts Sci. Mem.* **2**:73-81.

De Witt, B. 1820. *A Catalogue of Minerals Contained in the Cabinet of the Late Benjamin De Witt.* Albany, N.Y.: G. J. Loomis, 108p.

Gibbs, G. 1819. On the Tourmalines and Other Minerals Found at Chesterfield and Goshen, Massachusetts. *Am. Jour. Sci.* **1**:346-351.

Godon, S. 1809. Mineralogical Observations Made in the Environs of Boston, in the years 1807 and 1808. *Am. Acad. Arts Sci. Mem.* **3**:127-154, table.

Lea, Isaac. 1818. An Account of the Minerals at Present Known to Exist in the Vicinity of Philadelphia. *Phila. Acad. Nat. Sci. Jour.* **1**:462-482.

Phillips, W. 1818. *An Elementary Introduction to the Knowledge of Mineralogy.* New York: Collins, x, xxxiv, (10), 246, (10)p.

Seybert, A. 1800. A Catalogue of Some American Minerals, which Are Found in Different Parts of the United States. *Phila. Med. Mus.* **5**:152-159; 256-268.

Silliman, B. 1810. Sketch of the Mineralogy of the Town of New Haven. *Connecticut Acad. Arts Sci. Mem.* **1**:83-96.

Waterhouse, J. F. 1813. Description of Certain Minerals Found in the State of Massachusetts. *New England Jour. Med. Surg.* **2**:261-264.

MINERALOGY.

IS that science which teaches us the properties of mineral bodies, and by which we learn how to characterise, distinguish, and class them into a proper order.

INTRODUCTION.

MINERALOGY seems to have been in a manner coeval with the world. Precious stones of various kinds appear to have been well known among the Jews and Egyptians in the time of Moses; and even the most rude and barbarous nations appear to have had some knowledge of the ores of different metals. As the science is nearly allied to chemistry, it is probable that the improvements both in chemistry and mineralogy have nearly kept pace with each other; and indeed it is but of late, since the principles of chemistry were well understood, that mineralogy has been advanced to any degree of perfection. The best way of studying mineralogy, therefore, is by applying chemistry to it; and not contenting ourselves merely with inspecting the outsides of bodies, but decompounding them according to the rules of chemistry. This method has been brought to the greatest perfection by Mr Pott of Berlin, and after him by Mr Cronstedt of Sweden. To obtain this end, chemical experiments in the large way are without doubt necessary: but as a great deal of the mineral kingdom has already been examined in this manner, we do not need to repeat all those experiments in their whole extent, unless some new and particular phenomena should discover themselves in those things we are examining; else the tediousness of those processes might discourage some from going farther, and take up much of the time of others that might be better employed. An easier way may therefore be adopted, which even for the most part is sufficient, and which, though made in miniature, is as scientifical as the common manner of proceeding in the laboratories, since it imitates that, and is founded upon the same principles. This consists in making the experiments upon a piece of charcoal with the concentrated flame of a candle directed through a blow-pipe. The heat occasioned by this is very intense; and the mineral bodies may here be burnt, calcined, melted and scorified, &c. as well as in any great works.

For a description of the blow-pipe, the method of using it, the proper fluxes to be employed, and the different subjects of examination to which that instrument is adapted, see the article *Blow-pipe*, where all those particulars are concisely detailed. It may not be improper here, however, to resume those details at greater length; avoiding, at the same time, all unnecessary repetitions. After which we shall exhibit a scientific arrangement of the mineral kingdom according to the most approved system.

PART I. EXPERIMENTAL MINERALOGY; with a DESCRIPTION of the NECESSARY APPARATUS (A).

SECT. I. *Of experiments upon Earths and Stones.*

WHEN any of these substances are to be tried, we must not begin immediately with the blow-pipe; but some preliminary experiments ought to go before, by which those in the fire may afterwards be directed. For instance, a stone is not always homogenous, or of the same kind throughout, although it may appear to the eye to be so. A magnifying glass is therefore necessary to discover the heterogeneous particles, if there be any; and these ought to be separated, and every part tried by itself, that the effects of two different things, examined together, may not be attributed to one alone. This might happen with some of the finer micæ, which are now and then found mixed with small particles of quartz, scarcely to be perceived by the eye. The trapp (in German *schwartzstein*) is also sometimes mixed with very fine particles of feltspar *(spatam scintillans)* or of calcareous spar, &c. After this experiment, the hardness of the stone in question must be tried with steel. The flint and garnets are commonly known to strike fire with steel; but there are also other stones, which, though very seldom, are found so hard as likewise to strike fire. There is a kind of trapp of that hardness, in which no particles of feltspar are to be seen. Coloured glasses resemble true gems; but as they are very soft in proportion to these, they are easily discovered by means of the file. The common quartz-crystals are harder than coloured glasses, but softer than the gems. The loadstone discovers the presence of iron, when it is not mixed in too small a quantity in the stone, and often before the stone is roasted. Some kinds of hæmatites, and particularly the cærulescens, greatly resemble some other iron ores; but this distinguishes itself from them by a red colour when pounded, the others giving a blackish powder, and so forth.

The management of the *Blow-pipe* has been described under that article; but a few particulars may be here recapitulated, or added.

The candle ought to be snuffed often, but so that the top of the wick may retain some fat in it, because the flame is not hot enough when the wick is almost burnt to ashes; but only the top must be snuffed off, because a low wick gives too small a flame. The blue flame is the hottest; this ought, therefore, to be forced

---

(A) From Engestrom's *Treatise on the Blow-Pipe*, and Magellan's *Description of Pocket-Laboratories*, &c. subjoined to the English Translation of *Cronstedt's Mineralogy*, 2d edit. in 2 vols. Dilly.

forced out when a great heat is required, and only the point of the flame must be directed upon the subject which is to be essayed. M. Magellan recommends, as being most cleanly and convenient, that the candle be made of wax, and the wick should be thicker than ordinary. Its upper end must be bended towards the matter intended to be heated, and the stream of air must be directed along the surface of the bended part, so as not absolutely to touch it.

The piece of charcoal made use of in these experiments must not be of a disposition to crack. If this should happen, it must gradually be heated until it does not crack any more, before any assay is made upon it. If this be not attended to, but the assay made immediately with a strong flame, small pieces of it will split off in the face and eyes of the assayer, and often throw along with them the matter that was to be assayed. Charcoal which is too much burnt consumes too quick during the experiment, leaving small holes in it, wherein the matter to be tried may be lost; and charcoal that is burnt too little, catches flame from the candle, burning by itself like a piece of wood, which likewise hinders the process.

Of those things that are to be assayed, only a small piece must be broken off for that purpose, not bigger than that the flame of the candle may be able to act upon it at once, if required; which is sometimes necessary, as, when the matter requires to be made red hot throughout, the piece ought to be broken as thin as possible, at least the edges; the advantage of which is obvious, the fire having then more influence upon the subject, and the experiment being more quickly made.

Some of the mineral bodies are very difficult to be kept steady upon the charcoal during the experiment, before they are made red hot; because, as soon as the flame begins to act upon them, they split asunder with violence, and are dispersed. Such often are those which are of a soft consistence or a particular figure, and which preserve the same figure in however minute particles they are broken; for instance, the calcareous spar, the sparry gypsum, sparry fluor, white sparry lead-ore, the potters ore, the tessellated mock-lead or blende, &c. even all the common fluors which have no determinate figure. These not being so compact as common hard stones, when the flame is immediately urged upon them, the heat forces itself through and into their clefts or pores, and causes this violent expansion and dispersion. Many of the clays are likewise apt to crack in the fire, which may be for the most part ascribed to the humidity, of which they always retain a portion.

The only way of preventing this inconvenience is to heat the body as slowly as possible. It is best, first of all, to heat that place of the charcoal where the piece is intended to be put on; and afterwards lay it thereon; a little crackling will then ensue, but commonly of no great consequence. After that the flame is to be blown very slowly towards it, in the beginning not directly upon, but somewhat above it, and so approaching nearer and nearer with the flame until it become red hot. This will do for the most part; but there are nevertheless some, which, notwithstanding all the precautions, it is almost impossible to keep on the charcoal. Thus the fluors are generally the most difficult; and as one of their principal characters is discovered by their effects in the fire *per se*, they ought necessarily to be tried that way. To this purpose, it is best to make a little hole in the charcoal to put the fluor in, and then to put another piece of charcoal as a covering upon this, leaving only a small opening for the flame to enter. As this stone will nevertheless split and fly about, a larger piece thereof than is before-mentioned must be taken, in order to have at least something of it left.

But if the experiment is to be made upon a stone whose effects one does not want to see in the fire *per se*, but rather with fluxes, then a piece of it ought to be forced down into melted borax, when always some part of it will remain in the borax, notwithstanding the greatest part may sometimes fly away by cracking.

1. *Of substances to be tried in the fire* per se. As the stones undergo great alterations when exposed to the fire by themselves, whereby some of their characteristicks, and often the most principal, are discovered, they ought first to be tried that way, observing what has been said before concerning the quantity of matter, direction of the fire, &c. The following are generally the results of this experiment.

Calcareous earth or *stone*, when it is pure, does not melt by itself, but becomes white and friable, so as to break freely between the fingers; and, if suffered to cool, and then mixed with water, it becomes hot, just like common quick-lime. As in these experiments only very small pieces are used, this last effect is best discovered by putting the proof on the outside of the hand, with a drop of water to it, when instantly a very quick heat is felt on the skin. When the calcareous substance is mixed with the vitriolic acid, as in gypsum, or with a clay, as in marle, it commonly melts by itself, yet, more or less difficultly in proportion to the differences of the mixtures. Gypsum produces generally a white, and marle a grey, glass or flag. When there is any iron in it, as a white iron ore, it becomes dark, and sometimes quite black, &c.

The *siliciæ* never melt alone, but become generally more brittle after being burnt. Such of them as are coloured become colourless, and the sooner when it does not arise from any contained metal; for instance, the topazes, amethists, &c. some of the precious stones, however, excepted: And such as are mixed with a quantity of iron grow dark in the fire, as some of the jaspers, &c.

Garnets melt always into a black flag, and sometimes so easily that they may be brought into a round globule upon the charcoal.

The *argillaceæ*, when pure, never melt, but become white and hard. The same effects follow when they are mixed with phlogiston. Thus the soap-rock is easily cut with the knife; but being burnt it cuts glass, and would strike fire with the steel, if as large a piece as is necessary for that purpose could be tried in this way. The soap-rocks are sometimes found of a dark brown and nearly black colour, but nevertheless become quite white in the fire like a piece of China ware. However, care must be taken not to urge the flame from the top of the wick, there being for the most part a footy smoke, which commonly will darken all that it touches; and if this is not observed, a mistake in the experiment might easily happen. But if

On Earths and Stones.

it is mixed with iron, as it is sometimes found, it does not so easily part with its dark colour. The argillaceæ when mixed with lime melt by themselves, as above-mentioned. When mixed with iron, as in the boles, they grow dark or black; and if the iron is not in too great a quantity, they melt alone into a dark slag; the same happens when they are mixed with iron and a little of the vitriolic acid, as in the common clay, &c.

Mica and *asbestos* become somewhat hard and brittle in the fire, and are more or less refractory, though they give some marks of fusibility.

The *fluors* discover one of their chief characteristics by giving a light like phosphorus in the dark, when they are slowly heated; but lose this property, as well as their colour, as soon as they are made red hot.—They commonly melt in the fire into a white opaque slag, though some of them not very easily.

Some sorts of the *zeolites* melt easily, and foam in the fire, sometimes nearly as much as borax, and become a frothy slag, &c.

A great many of those mineral bodies which are impregnated with iron, as the *boles*, and some of the white iron ores, &c. as well as some of the other iron ores, viz. the bloodstone, are not attracted by the loadstone before they have been thoroughly roasted, &c.

2. *Of substances heated with fluxes.* After the mineral bodies have been tried in the fire by themselves, they ought to be heated with fluxes to discover if they can be melted or not, and some other phenomena attending this operation. For this purpose, three different kinds of salts are used as fluxes, viz. sal sodæ, borax, and sal fusible microsmicum; (see the article Blow-Pipe).

The *sal sodæ* is, however, not much used in these small experiments, its effects upon the charcoal rendering it for the most part unfit for it; because, as soon as the flame begins to act upon it, it melts instantly, and is almost wholly absorbed by the charcoal. When this salt is employed to make any experiment, a very little quantity is wanted at once, viz. about the cubical contents of an eighth part of an inch, more or less. This is laid upon the charcoal, and the flame blown on it with the blow-pipe; but as this salt commonly is in form of a powder, it is necessary to go on very gently, that the force of the flame may not disperse the minute particles of the salt. As soon as it begins to melt, it runs along on the charcoal, almost like melted tallow; and when cold, it is a glassy matter of an opaque dull colour spread on the coal. The moment it is melted, the matter which is to be tried ought to be put into it, because otherwise the greatest part of the salt will be soaked into the charcoal, and too little of it left for the intended purpose. The flame ought then to be directed on the matter itself; and if the salt spreads too much about, leaving the proof almost alone, it may be brought to it again by blowing the flame on its extremities, and directing it towards the subject of the experiment. In the assays made with this salt, it is true, we may find whether the mineral bodies which are melted with it have been dissolved by it or not: but we cannot tell with any certitude whether this is done hastily and with force, or gently and slow; nor whether a less or a greater part of the matter has been dissolved: neither can it be well distinguished if the water has imparted any weak tincture to the flux, because this salt always bubbles upon the charcoal during the experiment, nor is it clear when cool; so that scarcely any colour, except it be a very deep one, can be discovered, although it may sometimes be coloured by the matter that has been tried.

The following earths are entirely *soluble* in this flux with effervescence: Agate; chalcedony; carnelian; Turkey stone †, *(cos Turcica)*; fluor mineralis †; onyx; opal; quartz; common flint; ponderous spar. The following are *divisible* in it with or without effervescence, but not entirely soluble: Amianthus; asbestus; basaltes; chrysolite ‡; granate ‡; hornblende; jasper; marlstone; mica; the mineral of alum from Tolfa; petrosilex; aluminous slate and roof slate from Helsingia; emeralds; steatites; common flint; schoerl; talc; trapp; tripoli; tourmalin. And the following are neither fusible nor divisible in it: Diamond; hyacinth; ruby; sapphire; topaz.

The other two salts, viz. borax and the *sal microsmicum*, are very well adapted to these experiments, because they may by the flame be brought to a clear uncoloured and transparent glass; and as they have no attraction to the charcoal, they keep themselves always upon it in a round globular form. The sal fusible microcosmicum § is very scarce, and perhaps not to be met with in the shops; it is made of urine.

§ See Chemistry, 905, 906.

The following earths are soluble in borax, with more or less *effervescence*: Fluor mineralis †; marle; mica †; the mineral of alum from Tolfa; aluminous slate, and roof-slate from Helsingia †; ponderous spar; schoerl; talc †; tourmalin. And the following *without* effervescence; Agate; diamond; amianthus; asbestus; basaltes; chalcedony; carnelian; chrysolite; cos turcica; granate; hyacinth *; jasper; lapis ponderosus; onyx; opal; petro-silex; quartz *; ruby; sapphire; common flint *; steatite; trapp; trippel, or tripoli; topaz; zeolite; hydrophanes.

In the microcosmic salt, the following are soluble with more or less effervescence: Basaltes †; turkey stone ‡; fluor mineralis †; marle; mica; the mineral of alum from Tolfa; schistus aluminaris, schistus tegulalis from Helsingia †; schoerl; spathum ponderosum; tourmalin †; lapis ponderosus. And the following without visible effervescence: Agate; diamond; amianthus; asbestus; chalcedony; carnelian; chrysolite; granate; hyacinth; jasper; onyx ||; opal; petrosilex; quartz ||; ruby; sapphire; common flint ||; emerald; talc; topaz; trapp; trippel; zeolite; hornblend; hydrophanes; lithomarga; steatites.

Calcareous earth, ponderous spar, gypsum, and other additaments, often assist the solution, as well in the microcosmic salt as in borax. To which it is necessary to add, that in order to observe the effervescence properly, the matter added to the flux should be in the form of a small particle rather than in fine powder; because in this last there is always air between the particles, which being afterwards driven off by the heat afford the appearance of a kind of effervescence (A).

The

(A) In the above lists, the articles marked † effervesce very little; those marked ‡ not at all; those marked * require a larger quantity of the flux and a longer continuance of heat than the rest; those marked || are more difficultly dissolved than the others.

§ See Chemistry, no. 905, 906.

The quantity of those two salts required for an experiment is almost the same as the *sal sodæ*; but as the former are crystallised, and consequently include a great deal of water, particularly the borax, their bulk is considerably reduced when melted, and therefore a little more of them may be taken than the before mentioned quantity.

Both those salts, especially the borax, when exposed to the flame of the blow-pipe, bubble very much and foam before they melt to a clear glass, which for the most part depends on the water they contain. And as this would hinder the assayer from making due observations on the phenomena of the experiment, the salt which is to be used must first be brought to a clear glass before it can serve as a flux; it must therefore be kept in the fire until it become so transparent that the cracks in the charcoal may be seen through it. This done, whatsoever is to be tried is put to it, and the fire continued.

Here it is to be observed, that for the assays made with any of these two fluxes on mineral bodies may larger pieces must be taken that altogether they, no keep a globular form upon the charcoal; because it may then be better distinguished in what manner the flux acts upon the matter during the experiment. If this be not observed, the flux, communicating itself with every point of the surface of the mineral body, spreads all over it, and keeps the form of this last, which commonly is flat, and by that means hinders the operator observing all the phenomena which may happen. Besides, the flux being in too small a quantity in proportion to the body to be tried, will be too weak to act with all its force upon it. The best proportion therefore is about a third part of the mineral body to the flux; and as the quantity of the flux abovementioned makes a globe of a due size in regard to the greatest heat that is possible to procure in these experiments, so the size of the mineral body must be a third part less here than when it is to be tried in the fire by itself.

The *sal sodæ*, as has been already observed, is not of much use in these experiments; nor has it any particular qualities in preference to the two last mentioned salts, except that it dissolves the zeolites easier than they do.

The microcosmic salt shows almost the same effects in the fire as the borax, only differing from it in a very few circumstances; of which one of the principal is, that, when melted with manganese, it becomes of a crimson hue instead of a jacinth colour, which borax takes. This salt is, however, for its scarcity still very little in use, borax alone being that which is commonly employed. Whenever a mineral body is melted with any of these two last mentioned salts, in the manner already described, it is easily seen, Whether it quickly dissolves; in which case an effervescence arises, that lasts till the whole be dissolved: Whether the solution be slowly performed; in which case few and small bubbles only rise from the matter: or, Whether it can be dissolved at all; because if not, it is observed only to turn round in the flux, without the least bubble, and the edges look as sharp as they were before.

In order farther to illustrate what has been said about these experiments, we shall give a few examples of the effects of borax upon the mineral bodies.—The *calca-reous* substances, and all those stones which contain any thing of lime in their composition, dissolve readily and with effervescence in the borax. The effervescence is the more violent the greater the portion of lime contained in the stone. This cause, however, is not the only one in the gypsum, because both the constituents of this do readily mix with the borax, and therefore a greater effervescence arises in melting gypsum with the borax than lime alone.—The *siliceæ* do not dissolve; some few excepted which contain a quantity of iron.— The *argillaceæ*, when pure, are not acted upon by the borax: but when they are mixed with some heterogeneous bodies, they are dissolved, though very slowly; such are, for instance, the stone-marrow, the common clay, &c.

The *granates*, *zeolites*, and *trapp*, dissolve but slowly. The *fluors*, *asbestinæ*, and *micaceæ*, dissolve for the most part very easily; and so forth.—Some of these bodies melt to a colourless transparent glass with the borax; for instance, the calcareous substances when pure, the fluors, some of the zeolites, &c. Others tinge the borax with a green transparent colour, viz. the granates, trapp, some of the argillaceæ, and some of the micaceæ and asbestinæ. This green has its origin partly from a small portion of iron which the granates particularly contain, and partly from phlogiston.

Borax can only dissolve a certain quantity of the mineral body proportional to its own. Of the calcareous kind it dissolves a vast quantity; but turns at last, when too much has been added, from a clear transparent to a white opaque slag. When the quantity of the calcareous matter exceeds but little in proportion, the glass looks very clear as long as it remains hot: but as soon as it begins to cool, a white half opaque cloud is seen to arise from the bottom, which spreads over the third, half, or more of the glass globe, in proportion to the quantity of calcareous matter; but the glass or slag is nevertheless shining, and of a glassy texture when broken. If more of this matter be added, the cloud rises quicker and is more opaque, and so by degrees till the slag becomes quite milk white. It is then no more of a shining, but rather dry appearance, on the surface; is very brittle, and of a grained texture when broken.

Sect. II. *Of Experiments upon Metals and Ores.*

What has been hitherto said relates only to the *stones* and *earths:* We shall now proceed to describe the manner of examining *metals* and *ores*. An exact knowledge and nicety of procedure are so much the more necessary here, as the metals are often so disguised in their ores, as to be very difficultly known by their external appearance, and liable sometimes to be mistaken one for the other: Some of the cobalt ores, for instance, resemble much the *pyrites arsenicalis;* there are also some iron and lead ores, which are nearly like one another, &c.

As the ores generally consist of metals mineralised with sulphur or arsenic, or sometimes both together, they ought first to be exposed to the fire by themselves, in order not only to determine with which of these they are mineralised, but also to set them free from those volatile mineralising bodies: This serves instead of calcination, by which they are prepared for further assays.

Here

Here it must be repeated, that whenever any metal or fusible ore is to be tried, a little concavity must be made in that place of the charcoal where the matter is to be put; because, as soon as it is melted, it forms itself into a globular figure, and might then roll from the charcoal, if its surface was plain; but when borax is put to it, this inconvenience is not so much to be feared.

Whenever an ore is to be tried, a small bit being broke off for the purpose, it is laid upon the charcoal, and the flame blown on it slowly. Then the sulphur or arsenic begins to part from it in form of smoke: these are easily distinguished from one another by their smell; that of sulphur being sufficiently known, and the arsenic smelling like garlick. The flame ought to be blown very gently as long as any smoke is seen to part from the ore; but after that, the heat must be augmented by degrees, in order to make the calcination as perfect as possible. If the heat be applied very strongly from the beginning upon an ore that contains much sulphur or arsenic, the ore will presently melt, and yet lose very little of its mineralising bodies, by that means rendering the calcination very imperfect. It is, however, impossible to calcine the ores in this manner to the utmost perfection, which is easily seen in the following instance, *viz.* in melting down a calcined potter's ore with borax, it will be found to bubble upon the coal, which depends on the sulphur which is still left, the vitriolic acid of this uniting with the borax, and causing this motion. However, lead in its metallic form, melted in this manner, bubbles upon the charcoal, if any sulphur remains in it. But as the lead as well as some of the other metals, may raise bubbles upon the charcoal, although they are quite free from the sulphur, only by the flames being forced too violently on it, these phenomena ought not to be confounded with each other.

The ores being thus calcined, the metals contained in them may be discovered, either by being melted alone or with fluxes; when they show themselves either in their pure metallic state, or by tinging the flag with a colour peculiar to each of them. In these experiments it is not to be expected that the quantity of metal contained in the ore should be exactly determined; this must be done in larger laboratories. This cannot, however, be looked upon as any defect, since it is sufficient for a mineralogist only to find out what sort of metal is contained in the ore. There is another circumstance, which is a more real defect in the miniature laboratories, which is, that some ores are not at all capable of being tried by so small an apparatus; for instance, the gold ore called *pyrites aureus*, which consists of gold, iron, and sulphur. The greatest quantity of gold which this ore contains is about one ounce, or one ounce and an half, out of 100 pounds of the ore, the rest being iron and sulphur: and as only a very small bit is allowed for these experiments, the gold contained therein can hardly be discerned by the eye, even if it could be extracted; but it goes along with the iron in the flag, this last metal being in so large a quantity in proportion to the other, and both of them having an attraction for each other.

The blendes and black-jacks, which are mineral zinc ores, containing zinc, sulphur, and iron, cannot be tried this way, because they cannot be perfectly calcined, and besides the zinc flies off when the iron scorifies. Neither can those blendes, which contain silver or gold mineralised with them, be tried in this manner, which is particularly owing to the imperfect calcination. Nor are the quicksilver ores fit for these experiments: the volatility of that semimetal making it impossible to bring it out of the poorer sort of ores: and the rich ores, which sweat out the quicksilver when kept close in the hand, not wanting any of these assays, &c. These ores ought to be assayed in larger quantities, and even with such other methods as cannot be applied upon a piece of charcoal.

Some of the rich silver ores are easily tried: for instance, *minera argenti vitrea*, commonly called *silverglass*, which consists only of silver and sulphur. When this ore is exposed to the flame, it melts instantly, and the sulphur goes away in fume, leaving the silver pure upon the charcoal in a globular form. If this silver should happen to be of a dirty appearance, which often is the case, then it must be melted anew with a very little borax; and after it has been kept in fusion for a minute or two, so as to be perfectly melted and red-hot, the proof is suffered to cool: it may then be taken off the coal; and being laid upon the steel-plate †, the silver is separated from the flag by one or two strokes of the hammer †. Here the use of the brass ring† is manifest; for this ought first to be placed upon the plate, to hinder the proof from flying off by the violence of the stroke, which otherwise would happen. The silver is then found inclosed in the flag of a globular form, and quite shining, as if it was polished. When a large quantity of silver is contained in a lead ore, *viz.* in a potter's ore, it can likewise be discovered through the use of the blow-pipe, of which more will be mentioned hereafter.

Tin may be melted out of the pure tin ores in its metallic state. Some of these ores melt very easily, and yield their metal in quantity, if only exposed to the fire by themselves: but others are more refractory; and as these melt very slowly, the tin, which sweats out in form of very small globules, is instantly burnt to ashes before these globules have time to unite in order to compose a larger globe, which, might be seen by the eye, and not so soon destroyed by the fire; it is therefore necessary to add a little borax to these from the beginning, and then to blow the flame violently at the proof. The borax does here preserve the metal from being too soon calcined, and even contributes to the readier collecting of the small metalic particles, which soon are seen to form themselves into a globule of metallic tin at the bottom of the whole mass, nearest to the charcoal. As soon as so much of the metallic tin is produced as is sufficient to convince the operator of its presence, the fire ought to be discontinued, though the whole of the ore be not yet melted; because the whole of this kind of ore can be seldom or never reduced into metal by means of these experiments, a great proportion being always calcined: and if the fire is continued too long, perhaps even the metal already reduced may likewise be burnt to ashes; for the tin is very soon deprived of its metallic state by the fire.

Most part of the lead ores may be reduced to a metallic state upon the charcoal. The *minerae plumbi calciformes*, which are pure, are easily melted into lead; but

†See the article Blow-Pipe and Plate XCIX.

MINERALOGY.

but such of them as are mixed with an *ochra ferri*, or any kind of earth, as clay, lime, &c. yield very little of lead, and even nothing at all, if the heterogenea are combined in any large quantity: this happens even with the *minera plumbi calciformis arsenico mixta*. These therefore are not to be tried but in larger laboratories. However, every mineral body suspected to contain any metallic substance may be tried by the blow-pipe, so as to give sufficient proofs whether it contain any or not, by its effects being different from those of the stones or earths, &c.

The *mineræ plumbi mineralisatæ* leave the lead in a metallic form, if not too large a quantity of iron is mixed with it. For example, when a tessellated or steel-grained lead ore is exposed to the flame, its sulphur, and even the arsenic if there be any, begins to fume, and the ore itself immediately to melt into a globular form; the rest of the sulphur continues then to fly off, if the flame be blown slowly upon the mass; but, on the contrary, very little of the sulphur will go off, if the flame be forced violently on it: in this case, it rather happens that the lead itself crackles and dissipates, throwing about very minute metallic particles. The sulphur being driven out as much as possible, which is known by finding no sulphureous vapour in smelling at the proof, the whole is suffered to cool, and then a globule of metallic lead will be left upon the coal. If any iron is contained in the lead-ore, the lead, which is melted out of it, is not of a metallic shining, but rather of a black and uneven, surface: a little borax must in this case be melted with it, and as soon as no bubble is seen to rise any longer from the metal into the borax, the fire must be discontinued: when the mass is grown cold, the iron will be found scorified with the borax, and the lead left pure and of a shining colour.

Borax does not scorify the lead in these small experiments when it is pure: if the flame is forced with a violence on it, a bubbling will ensue, resembling that which is observed when borax dissolves a body melted with it; but when the fire ceases, the flag will be perfectly clear and transparent, and a quantity of very minute particles of lead will be seen spread about the borax, which have been torn off from the mass during the bubbling.

If such a lead ore is rich in silver, this last metal may likewise be discovered by this experiment; because as the lead is volatile, it may be forced off, and the silver remain. To effect this, the lead, which is melted out of the ore, must be kept in constant fusion with a slow heat, that it may be consumed. This end will be sooner obtained, and the lead part quicker, if during the fusion the wind through the blow-pipe be directed immediately, though not forcibly, upon the melted mass itself, until it begin to cool; at which time the fire must be directed on it again. The lead, which is already in a volatilising state, will by this artifice be driven out in form of a subtil smoke; and by thus continuing by turns to melt the mass, and then to blow off the lead, as has been said, until no smoke is any longer perceived, the silver will at last be obtained pure. The same observation holds good here also, which was made about the gold, that, as none but very little bits of ores can be employed in these experiments, it will be difficult to extract the silver out of a poor ore: for some part of it will fly off with the lead, and what might be left is too small to be discerned by the eye. The silver, which by this means is obtained, is easily distinguished from lead by the following external marks, *viz.* that it must be red-hot before it can be melted: it cools sooner than lead; it has a silver colour; that is to say, brighter and whiter than lead: and is harder under the hammer.

The *mineræ cupri calciformes* (at least some of them), when not mixed with too much stone or earth, are easily reduced to copper with any flux; if the copper is found not to have its natural bright colour, it must be melted with a little borax, which purifies it. Some of these ores do not all discover their metal if not immediately melted with borax; the heterogenea contained in them hindering the fusion before these are scorified by the flux.

The grey copper ores, which only consist of copper and sulphur, are tried almost in the same manner as abovementioned. Being exposed to the flame by themselves, they will be found instantly to melt, and part of their sulphur to go off. The copper may afterwards be obtained in two ways: the one, by keeping the proof in fusion for about a minute, and afterwards suffering it to cool; when it will be found to have a dark and uneven appearance externally, but which after being broken discovers the metallic copper of a globular form in its centre, surrounded with a regulus, which still contains some sulphur and a portion of the metal: the other, by being melted with borax, which last way sometimes makes the metal appear sooner.

The *mineræ cupri pyritaceæ*, containing copper, sulphur, and iron, may be tried with the blow-pipe if they are not too poor. In these experiments the ore ought to be calcined, and after that the iron scorified. For this purpose a bit of the ore must be exposed to a slow flame, that as much of the sulphur as possible may part from it before it is melted, because the ore commonly melts very soon, and then the sulphur is more difficultly driven off. After being melted, it must be kept in fusion with a strong fire for about a minute, that a great part of the iron may be calcined; and after that, some borax must be added, which scorifies the iron, and turns with it to a black flag. If the ore is very rich, metallic copper will be had in the flag after the scorification. If the ore be of a moderate richness, the copper will still retain a little sulphur, and sometimes iron: the product will therefore be brittle, and must with great caution be separated from the flag, that it may not break into pieces; and if this product is afterwards treated in the same manner as before said, in speaking of the grey copper-ores, the metal will soon be produced. But if the ore is poor, the product after the first scorification must be brought into fusion, and afterwards melted with some fresh borax, in order to calcine and scorify the remaining portion of iron; after which it may be treated as mentioned in the preceding paragraph. The copper will in this last case be found in a very small globule.

The copper is not very easily scorified with this apparatus, when it is melted together with borax, unless it has first been exposed to the fire by itself for a while in order to be calcined. When only a little of this metal is dissolved, it instantly tinges the flag of a reddish

MINERALOGY.

On Metals and Ores.

difh brown colour, and moftly opaque; but as foon as this flag is kept in fufion for a little while, it becomes quite green and tranfparent: and thus the prefence of the copper may be difcovered by the colour, when it is concealed in heterogeneous bodies, fo as not to be difcovered by any other experiment.

If metallic copper is melted with borax by a flow fire, and only for a very little time, the glafs or flag becomes of a fine tranfparent blue or violet colour, inclining more or lefs to the green: but this colour is not properly owing to the copper, but it may rather be to its phlogifton; becaufe the fame colour is to be had in the fame manner from iron; and thefe glaffes, which are coloured with either of thofe two metals, foon lofe their colour if expofed to a ftrong fire, in which they become quite clear and colourlefs. Befides, if this glafs, tinged blue with the copper, is again melted with more of this metal, it becomes of a good green colour, which for a long time keeps unchanged in the fire.

The iron ores, when pure, can never be melted *per fe*, by the means of the blow-pipe alone; nor do they yield their metal when melted with fluxes; becaufe they require too ftrong a heat to be brought into fufion; and as both the ore and the metal itfelf very foon lofe their phlogifton in the fire, and cannot be fupplied with a fufficient quantity from the charcoal, fo likewife they are very foon calcined in the fire. This eafy calcination is alfo the reafon why the fluxes, for inftance borax, readily fcorify this ore, and even the metal itfelf. The iron lofes its phlogifton in the fire fooner than the copper, and is therefore more eafily fcorified.

The iron is, however, difcovered without much difficulty, although it were mixed but in a very fmall quantity with heterogeneous bodies. The ore, or thofe bodies which contain any large quantity of the metal, are all attracted by the loadftone, fome without any previous calcination, and others without having been roafted. When a clay is mixed with a little iron, it commonly melts by itfelf in the fire; but if this metal is contained in a limeftone; it does not promote the fufion, but gives the ftone a dark and fometimes a deep black colour, which always is the character of iron. A *minera ferri calciformis pura cryftallifata*, is commonly of a red colour: This being expofed to the flame, becomes quite black; and is then readily attracted by the loadftone, which it was not before. Befides thefe figns, the iron difcovers itfelf, by tinging the flag of a green tranfparent colour, inclining to brown, when only a little of the metal is fcorified; but as foon as any larger quantity thereof is diffolved in the flag, this becomes firft a blackifh brown, and afterwards quite black and opaque.

Bifmuth is known by its communicating a yellowifh brown colour to borax; and arfenic by its volatility and garlick fmell. Antimony, both in form of regulus and ore, is wholly volatile in the fire when it is not mixed with any other metal except arfenic; and is known by its particular fmell, eafier to be diftinguifhed when once known than defcribed. When the ore of antimony is melted upon the charcoal, it bubbles conftantly during its volatilifing.

Zinc ores are not eafily tried upon the coal; but the regulus of zinc expofed to the fire upon the charcoal burns with a beautiful blue flame, and forms itfelf almoft inftantly into white flowers, which are the common flowers of zinc.

Cobalt is particularly remarkable for giving to the glafs a blue colour, which is the zaffre or fmalt. To produce this, a piece of cobalt ore muft be calcined in the fire, and afterwards melted with borax. As foon as the glafs, during the fufion, from being clear, feems to grow opaque, it is a fign that it is already tinged a little; the fire is then to be difcontinued, and the operator muft take hold, with the nippers, of a little of the glafs, whilft yet hot, and draw it out flowly in the beginning, but afterwards very quick, before it cools, whereby a thread of the coloured glafs is procured, more or lefs thick, wherein the colour may eafier be feen than in a globular form. This thread melts eafily, if only put in the flame of the candle without the help of the blow-pipe.—If this glafs be melted again with more of the cobalt, and kept in fufion for a while, the colour becomes very deep; and thus the colour may be altered at pleafure.

When the cobalt ore is pure, or at leaft contains but little iron, a cobalt regulus is almoft inftantly produced in the borax during the fufion; but when it is mixed with a quantity of iron, this laft metal ought firft to be feparated, which is eafily performed fince it fcorifies fooner than the cobalt; therefore, as long as the flag retains any brown or black colour, it muft be feparated, and melted again with frefh borax, until it fhows the blue colour.

Nickel is very feldom to be had; and as its ores are feldom free from mixtures of other metals, it is very difficultly tried with the blow-pipe. However, when this femimetal is mixed with iron and cobalt, it is eafily freed from thefe heterogeneous metals, and reduced to a pure nickel regulus by means of fcorification with borax, becaufe both the iron and cobalt fooner fcorify than the nickel. The regulus of nickel itfelf is of a green colour when calcined: it requires a pretty ftrong fire before it melts, and tinges the borax with a hyacinth colour. Manganefe gives the fame colour to borax; but its other qualities are quite different, fo as not to be confounded with the nickel.

By means of the foregoing explanations, and thofe given under the article *Blow-Pipe*, any gentleman, who is a lover of this fcience, will be able, in an eafy manner, to amufe himfelf in difcovering the properties of thofe works of nature, with which the mineral kingdom furnifhes us; or more ufefully to employ himfelf by finding out what forts of ftones, earths, ores, &c. there are on his eftate, and to what economical purpofes they may be employed. The fcientific mineralift may, by examining into the properties and effects of the mineral bodies, difcover the natural relation thefe bodies ftand in to each other. and thereby furnifh himfelf with materials for eftablifhing a mineral fyftem, founded on fuch principles as Nature herfelf has laid down in them; and this in his own ftudy, without being forced to have recourfe to great laboratories, crucibles, furnaces, &c. which is attended with much trouble, and is the reafon why fo few can have an opportunity of gratifying their defire of knowledge in

this

MINERALOGY.

this part of natural history. Farther improvements of this apparatus may still be made by those who choose to bestow their attention upon it.

A great number of fluxes might, perhaps, be found out, whose effects might be different from those already in use, whereby more distinct characters of those mineral bodies might be discovered, which now either show ambiguous ones, or which it is almost impossible to try exactly with the blowpipe. Instead of the *sal sodæ*, some other salt might be discovered better adapted to these experiments. But it is very necessary not to make use of any other fluxes on the charcoal than such as have no attraction to it: if they, at the same time, be clear and transparent, when melted, as the borax and the *sal fusibile microcosmicum*, it is still better: however, the transparency and opacity are of no great consequence, if a substance be assayed only in order to discover its fusibility, without any attention to its colour; in which case, some metallic slag, perhaps, might be useful.

When such ores are to be reduced whose metals are very easily calcined, as tin, zinc, &c. it might perhaps be of service to add some phlogistic body, such as hard resin, since the charcoal cannot afford enough of it in the open fire of these assays. The manner of melting the volatile metals out of their ores *per descensum* might also, perhaps, be imitated: for instance, a hole might be made in the charcoal, wide above and very narrow at the bottom; a little piece of the ore being then laid at the upper end of the hole, and covered with some very small pieces of the charcoal, the flame must be directed on the top: the metal might, perhaps, by this method, run into the hole below, concealed from the violence of the fire, particularly if the ore is very fusible, &c.

The use of the apparatus above referred to, and which may be called a *pocket laboratory* (as the whole admits of being easily packed into a small case), is chiefly calculated for a travelling mineralist. But a person who always resides at one and the same place, may by some alteration make it more commodious to himself, and avoid the trouble of blowing with the mouth. For this purpose he may have the blow-pipe go through a hole in a table, and fixed underneath to a small pair of bellows with double bottoms, such as some of the glass-blowers use, and then nothing more is required than to move the bellows with the feet during the experiment; but in this case a lamp may be used instead of a candle. This method would be attended with a still greater advantage, if there were many such parts as *c*, fig. 13. the openings of which were of different dimensions: those parts might by means of a screw be fastened to the main body of the blow-pipe, and taken away at pleasure. The advantage of having these nozzles of different capacities at their ends, would be that of exciting a stronger or weaker blast as occasion might require. It would only be necessary to observe, that in proportion as the opening or nozzle of the pipe is enlarged, the quantity of the flame must be augmented by a thicker wick in the lamp, and the force of blowing encreased by means of weights laid on the bellows; a much intenser heat would thus be produced by a pipe of a considerable opening at the end, by which the experiments must undoubtedly be carried farther than the common blow-pipe.

Portable Apparatus.

A traveller, who has seldom an opportunity of carrying many things along with him, may very well be contented with this laboratory and its apparatus, which are sufficient for most part of such experiments as can be made on a journey. There are, however, other things very useful to have at hand on a journey, which ought to make a separate part of a portable laboratory, if the manner of travelling does not oppose it: this consists of a little box including the different acids, and one or two matrasses, in order to try the mineral bodies in liquid menstrua if required.

These acids are, the acid of nitre, of vitriol, and of common salt. Most of the stones and earths are attacked, at least in some degree, by the acids; but the calcareous are the easiest of all to be dissolved by them, which is accounted for by their calcareous properties. The acid of nitre is that which is most used in these experiments; it dissolves the limestone, when pure, perfectly, with a violent effervescence, and the solution becomes clear: when the limestone enters into some other body, it is nevertheless discovered by this acid, through a greater or less effervescence in proportion to the quantity of the calcareous particles, unless there are so few as to be almost concealed from the acid by the heterogeneous ones. In this manner a calcareous body, which sometimes nearly resembles a siliceous or argillaceous one, may be known from these latter, without the help of the blow-pipe, only by pouring one or two drops of this acid upon the subject; which is very convenient when there is no opportunity nor time of using this instrument.

The gypsa, which consist of lime and the vitriolic acid, are not in the least attacked by the acid of nitre, if they contain a sufficient quantity of their own acid; because the vitriolic acid has a stronger attraction to the lime than the acid of nitre: but if the calcareous substance is not perfectly saturated with the acid of vitriol, then an effervescence arises with the acid of nitre, more or less in proportion to the want of the vitriolic acid. These circumstances are often very essential in distinguishing the *calcarea* and *gypsa* from one another.

The acid of nitre is likewise necessary in trying the zeolites, of which some species have the singular effect to dissolve with effervescence in the abovementioned acid; and within a quarter of an hour, or even sometimes not until several hours after, to change the whole solution into a clear jelly, of so firm a consistence, that the glass wherein it is contained may be reversed without its falling out.

If any mineral body is tried in this menstruum, and only a small quantity is suspected to be dissolved, though it was impossible to distinguish it with the eye during the solution, it can be easily discovered by adding to it *ad saturitatem* a clear solution of the alkali, when the dissolved part will be precipitated, and fall to the bottom. For this purpose the *sal sodæ* may be very useful.

The acid of nitre will suffice for making experiments upon stones and earths; but if the experiments are to be extended to the metals, the other two acids are also necessary.

Another instrument is likewise necessary to a complete

Vol. XII. I

complete Pocket Laboratory, viz. a washing-trough (fig. 21.), in which the mineral bodies, and particularly the ores, may be separated from each other, and from the adherent rock, by means of water. This trough is very common in laboratories, and is used of different sizes; but here only one is required of a moderate size, such as 12 inches and a half long, three inches broad at the one end and one inch and a half at the other end, sloping down from the sides and the broad end to the bottom, where it is three quarters of an inch deep. It may, however, be made of much smaller dimensions. It is commonly made of wood, which ought to be chosen smooth, hard, and compact, wherein are no pores in which the minute grains of the pounded matter may conceal themselves. It is to be observed, that if any such matter is to be washed as is suspected to contain some native metal, such as silver or gold, a trough should be procured for this purpose of a very shallow slope; because the minute particles of the native metal have then more power to assemble together at the broad end, and separate from the other matter.

The management of this trough, or the manner of washing, consists in this: That when the matter is mixed with about three or four times its quantity of water in the trough, this is kept very loose between two fingers of the left hand, and some light strokes given on its broad end with the right, that it may move backwards and forwards; by which means the heaviest particles assemble at the broad and lower end, from which the lighter ones are to be separated by inclining the trough and pouring a little water on them. By repeating this process, all such particles as are of the same gravity may be collected together, and separated from those of different gravity, provided they were before equally pounded: though such as are of a clayey nature, are often very difficult to separate from the rest, which, however, is of no great consequence to a skilful and experienced washer. The washing process is very necessary, as there are often rich ores, and even native metals, found concealed in earths and sand in such minute particles as not to be discovered by any other means.

SECT. III. *Description of an Improved Portable Laboratory for assaying Minerals.*

THE chief pieces and implements of the portable laboratories are represented in Plate XCIX. at BLOW-PIPE, and in Plate CCCXIII. annexed to the present article.

I. The first contains those belonging to the *Dry* Laboratory, so called on account of its containing whatever is required to try all kinds of fossils in the dry way by fire, without any of the humid menstruums. They are made to pack in a box of the size of an octavo book, lined with green velvet, and covered with black fish-skin; the inside divided into different compartments, suited to the size, form, and number of the implements it is to contain. Of these the principal are described under BLOW-PIPE. We must here, however, add the following remarks and alterations of that instrument by Mr Magellan.

D and Q (fig. 13.) are the two pieces that form the blow-pipe, which is here represented entire. This very useful instrument has been considerably improved of late in England. The mouth-piece *a a* is made of ivory, to avoid the disagreeable sensation of having a piece of metal a long time between the teeth and lips, which, if not of silver or gold, may be very noxious to the operator; a circumstance that has been hardly noticed before.

1. If the mouth-piece *a a* be made of a round form, it cannot be held for any length of time between the teeth and lips, to blow through it, without straining the muscles of the mouth, which produces a painful sensation. It must, therefore, have such an external figure, as to adapt itself accurately to the lateral angles of the lips, having a flattish oval form externally, with two opposite corners to fit those internal angles of the mouth, when it is held between the lips, as may be seen in that represented in the figure.

2. The small globe *b b* is hollow, for receiving the moisture of the breath; and must be composed of two hemispheres, exactly screwing into one another in *b b*; the male-screw is to be in the lower part, and soldered on the crooked part Q of the tube Q D, at such a distance, that the inside end of the crooked tube be even with the edge of the hemisphere, as represented by the pointed lines in the figure. But the upper hemisphere is to be soldered at the end of the straight tube D. By these means, the moisture arising from the breath falls into the hollow of the lower hemisphere, where it is collected round the upper inside end of the crooked part Q of the blow-pipe, without being apt to fall into it.

3. The small nozzles, or hollow conical tubes, advised by Messrs Engestrom, Bergman, and others, are wrong in the principle; because the wind that passes from the mouth through such long cones loses its velocity by the lateral friction, as happens in hydraulic spouts; which, when formed in this manner, do never throw the fluid so far as when the fluid passes through a hole of the same diameter, made in a thin plate of a little metallic cap that screws at the end of the large pipe. It is on this account that the little cap *c* is employed, having a small hole in the thin plate, which serves as a cover to it; and there are several of these little caps, with holes of smaller and larger sizes, to be changed and applied whenever a flame is required to be more or less strong.

4. Another convenience of these little caps is, that even in case any moisture should escape falling into the hemisphere *b b*, and pass along with the wind through the crooked pipe Q, it never can arrive at nor obstruct the little hole of the cap *c*, there being room enough under the hole in the inside, where this moisture must be stopped till it is cleaned and wiped out.

The stream of air that is impelled by the blow-pipe (as seen in fig. 3.) upon the flame, must be constant and even, and must last as long as the experiment continues to require it. This labour will fatigue the lungs, unless an equable and uninterrupted inspiration can at the same time be continued. To succeed in this operation without inconvenience, some labour and practice are necessary, as already explained under the detached article.

Every assay ought always to begin by the exterior flame, which must be first directed upon the mass under examination; and, when its efficacy is well known, then the interior blue flame is to be employed.

After

MINERALOGY Plate CCCXIII

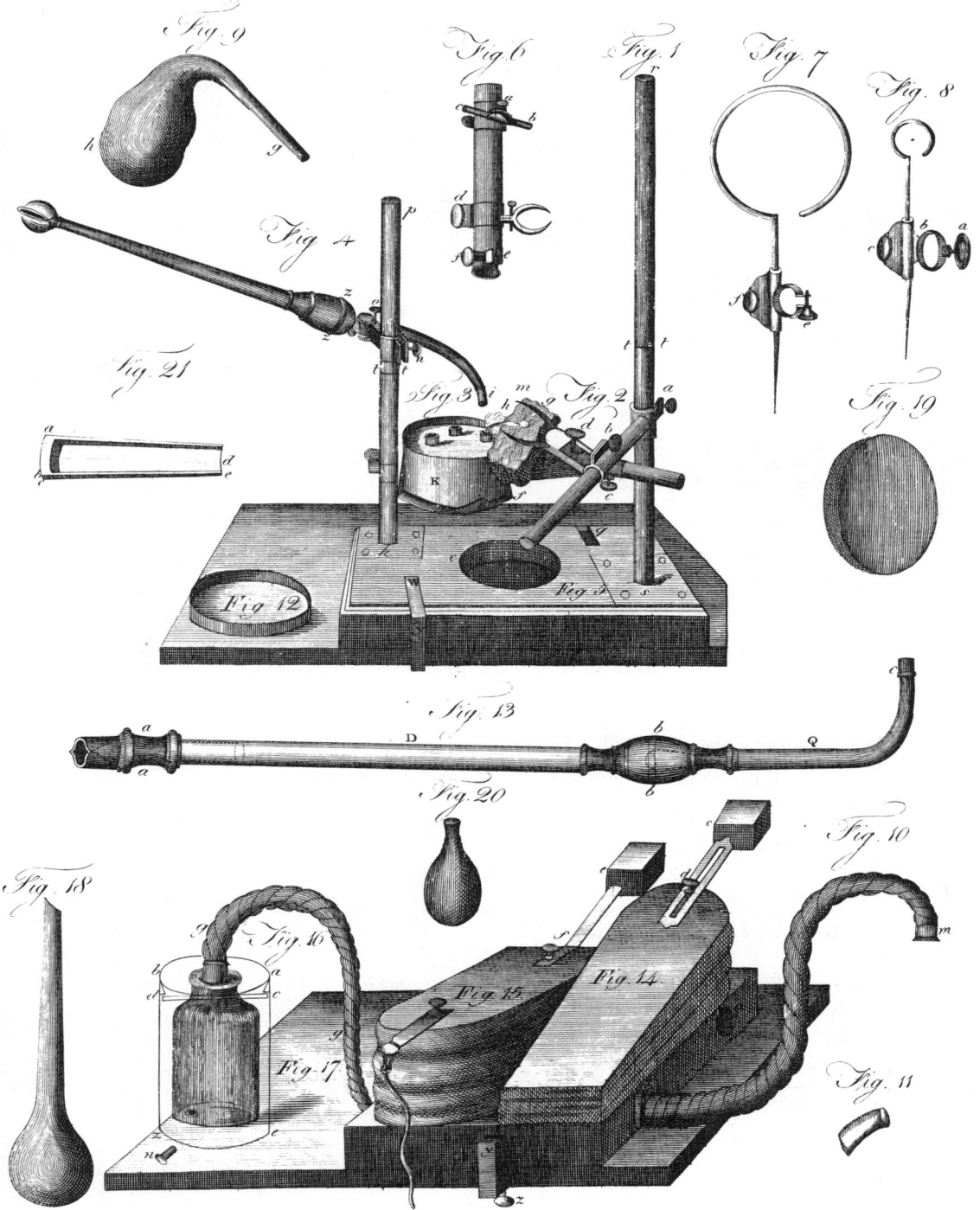

R. Scot & S. Allardice

After the ore is roasted, it is to be rounded upon the steel plate by the hammer; the particles being prevented from being dissipated by the ring H (fig. 9, Plate XCIX.), within which the pieces to be broken are to be put.

Among the apparatus, beside the particulars already mentioned, three phials are necessary, containing the required fluxes, viz. the *borax*, the *sal sodæ*, and *sal fusile microcosmicum*. Other useful particulars are, A small link of hard steel, to try the hardness or softness of mineral substances, and also to strike fire for lighting the candle when required: A piece of black flint, to serve as a touch-stone; (for being rubbed with any metal, if it be gold the marks will not be corroded by aqua fortis); and also to strike fire, when necessary, with the link of steel: An artificial load-stone, properly armed with iron, for the better preservation of its attractive power; (it serves to discover the ferrugineous particles of any ore after it has been roasted and powdered:) A triple magnifier, which, differently combined, produces seven magnifying powers, the better to distinguish the structure and metallic parts of ores, and the minute particles of native gold, whenever they contain that metal: A file, to try the hardness of stones and crystals, &c.: Some pieces of dry agaric or tinder, and small bits or splinters of wood tipped with brimstone, to serve as matches for lighting the candle; and various other little articles of use in these experiments.

II. For performing experiments in the *Humid Way*, the chief additional articles (and which must be kept in a separate case) consist of a collection of phials, containing the principal acids, tests, precipitants, and re-agents, both for examining mineral bodies by the humid way, and for analysing the various kinds of mineral waters. Those with acids and corrosive solutions have not only ground stoples, but also an external cap to each, ground over the stople, and secured downward by a bit of wax between both, in order to confine the corrosive and volatile fluids within. But those which contain mild fluid liquors have not such external caps: and those with dry inoffensive substances are only stopped with cork. Besides these phials, there are two smaller cylindrical ones, which serve to exhibit the changes of colour produced by some of the re-agents in those analytical assays. There are also two or three small matrasses, to hold the substances with their solvents over the fire; a small glass funnel, for pouring the fluids; a small porcelain mortar, with its pestle; one or two crucibles of the same substance; a small wooden trough to wash the ground ores; some glass sticks to stir up the fluid mixtures; and, finally, pieces of paper tinged red, yellow, and blue, by the tinctures of Fernambuc wood (commonly called Brasil wood), turmeric, and litmus, thickened with a little starch.

The following list contains the names of the various fluid tests and re-agents that are necessary for these assays. But the whole number being too large to be all contained in a portable case, every one may give the preference to those he likes best.

Portable Apparatus.

1. Concentrated vitriolic acid, whose specific gravity may be expressed in the outside.
2. Nitrous acid, purified by the nitrous solution of silver.
3. Concentrated marine acid, with its specific gravity.
4. Marine acid dephlogisticated.
5. Aqua regia for gold, viz. 2 nit. and 1 marine.
6. Aqua regia for platina viz. half marine and half nitrous acid.
7. Nitrous solution of silver.
8. Nitrous solution of mercury, made in the cold.
9. Muriatic solution of barytes.
10. Nitrous solution of lime.
11. Muriatic solution of lime.
12. Mercury in its metallic state.
13. Corrosive sublimate of mercury.
14. White arsenic.
15. Nitrous solution of silver.
16. Nitrous solution of copper.
17. Acid of sugar.
18. Liquor probatorius vini.
19. Hepar sulphuris.
20. Oil of tartar *per deliquium*.
21. Salt of tartar.
22. Caustic vegetable alkali.
23. Pearl-ashes.
24. Soap-makers ley.
25. Common salt.
26. Vitriolated argilla (alum.)
27. Vitriol of iron (copperas)
28. Nitrous solution of silver.
29. Acetous solution of lead.
30. Acetous solution of barytes.
31. Phlogisticated alkali by the Prussian blue.
32. Lime-Water.
33. Lime-water phlogisticated by the Prussian blue.
34. Caustic volatile alkali.
35. Mild volatile alkali (dry.)
36. Rectified spirit (alcohol.)
37. Æther.
38. Spirituous tincture of galls.

The following tests are very fit also for these assays, viz. 39. Spirituous solutions of soap: 40. Syrup of violets; 41 Tincture of litmus; 42. Tincture of Brasil wood; 43. Tincture of turmeric; 44. Oil of olives; 45. Oil of linseed: 46. Oil of turpentine; 47. Essential salt of wild-sorrel; 48. Hepar sulphuris: 49. Sugar of lead; 50. Solution of alum.

The method of applying the above tests of acids and re-agents may be seen in Bergman's treatises of the Analysis of Waters, and of Assaying by the Humid Way; in Kirwan's Elements of Mineralogy; in the Elements of Chemistry of Dijon; in the Memoirs of the same Academy; in Fourcroy's Lectures of Chemistry, &c.

III. The *Lamp-furnace* Laboratory, for experiments both by the *humid* and the *dry* way, is a very curious and useful, though small apparatus. It is an improvement of that which was contrived by M. de Morveau, in consequence of the information he received from his friend the president de Virly, who saw at Upsal how advantageously the late eminent professor Bergman availed himself of this convenience for many analytical processes in miniature, by the use of very small glass vessels about one inch diameter, and other implements of proportional size, for performing various chemical operations. (See the Dijon Memoirs for 1783. Part I. p. 171.)

There

MINERALOGY.

Portable Apparatus.

There can be no doubt but that whenever these processes are properly conducted, though in miniature, the lamp-furnace will prove amply sufficient to perform in a few minutes, and with very little expence, the various solutions, digestions, and distillations, which otherwise would require large vessels, stills, retorts, reverberatory furnaces, &c. to ascertain the component parts of natural bodies; though it is not always sufficient to ascertain their respective quantities. In this last case, operations must be performed in great laboratories, and on a large scale, at a considerable expence. But the substances are sometimes too valuable; as, for instance, when precious stones are examined; and of course the last way never can be attempted in such cases.

These small processes have likewise another advantage before noticed, which cannot be obtained in works at large. It consists in one's being able to observe the gradual progress of each operation; of easily retarding or urging it, as it may require; and of ascertaining at pleasure each step of every experiment, together with the phenomena attending the same.

Plate CCCXIII.

The lamp-furnace is mounted in a small parallelogram of mahogany, about six inches long and four wide, marked fig. 5. This is kept steady over the edge of a common table, by means of the metallic clamp wxv, which is fastened by the screw x. The pillar rs is screwed in a vertical position on the plate s, being about ten inches high: the other is screwed to the opposite corner, marked pk, and is only $7\frac{1}{4}$ inches long; both are composed of two halves, that screw at tt, to be easily packed up with all the implements in a case covered with black fish skin, and lined with green velvet, like the other laboratory already described.

The lamp k, fig. 3. is supported on the plate f, which has a ring l that runs in the column pk, and may be fixed by its screw l at the required height.— This lamp has three small pipes of different sizes, to receive as many wicks of different thickness, and to be filled with spirit of wine. By a similar method, a piece of charcoal is mounted and supported by the pliers or little forceps screwed to the arm ac, fig. 1. which has all the motions requisite for being fixed by means of proper screws, at a proper distance from the flame of the wick h. The blow-pipe, fig. 4. is, by a similar mechanism, mounted on the smaller column pq, at such a distance as to blow the flame hi to the piece of ore m, which is upon the charcoal gf.

Every thing being disposed in this manner, the operator blows through the mouth-piece of the blow-pipe, fig. 4. and remains with his hands free to make the changes and alterations he may think proper.—

[*N. B.* The large round cavity e in the middle of the parallelogram, fig. 5. is to receive the lamp k, fig. 3. when all the implements are packed up in their case of black fish skin: and the cover of the lamp is represented by fig. 12.]

But if the operator has the double bellows, fig. 14. and 15. he fixes them, at a due distance, to the same table by the brass clamp y. He then unscrews the blow-pipe at z z, joins the mouth m of the flexible tube of the hemisphere zz, passing each orifice thro' the leather tube fig. 11. and tying both ends with a waxed thin pack-thread. If he works with his foot on the pedal, the string of which is seen hanging from the end of the bellows, fig. 15. (and is always up, on account of the weight e), then the air is absorbed by the bellows fig. 15. from whence it is propelled by the motion of the foot on the pedal to the bellows, fig. 14. whose constant weight r drives it out through the flexible pipe, fig. 10. it of course enters the curbed part zzi of the blow-pipe, and drives the flame on the piece m of the ore, that is to be examined upon the charcoal.

[*N. B.* 1. This double bellows is packed up by itself in a mahogany case, about 9 inches long. $6\frac{1}{4}$ wide, and about $3\frac{1}{4}$ deep, outside measure. 2. The last blowing bellows, fig. 14. has an inside valve, which opens when the upper surface of it is at its greatest height; in order to let the superfluous air escape out, as it would otherwise issue with great velocity out of the tube, fig, 11. and spoil the operation.]

If the operator chooses to apply the vital or dephlogisticated air in his process, let him fill the glass jar b, fig. 17. with this air; and put it within the tube marked by $abze$, filled with water, fastening the neck of the jar within by a cross-board cd, which has a hole in it for that purpose: then introducing the two ends of the flexible hollow tube, fig. 16. both to the mouth of the jar and to the hole of the bellows fig. 15. he opens the hole m of the jar, that was stopped with the stopple n; the column of the water passes in through m, and forces up the vital air, which enters the bellows, and of course, by the alternate motion of the pedal, passes through the end of the blow-pipe, to urge the flame upon the piece of ore m, fig. 2. on the charcoal g. But the dephlogisticated air may be also received at the same time that it is produced, by tying the pipe, fig. 16. to the mouth of an earthen retort, or even of a glass retort well-coated, according to the method of Mr Willis, described in the Transactions of the Society of Arts, Vol. V. p. 96. This last consists in dissolving two ounces of borax in a pint of boiling water, and adding to the solution as much slacked lime as is necessary to form a thin paste. This glass retort is to be covered all over with it, by means of a painter's brush, and then suffered to dry. It must then be covered with a thin paste made of linseed oil and slacked lime, except the neck that enters into the receiver. In two or three days it will dry of itself; and the retort will then bear the greatest fire without cracking. Two ounces of good nitre being urged in the retort, by a good fire on a chafing-dish, will afford about 700 or 800 ounce-measures of dephlogisticated air.

To make any other kind of chemical assays, the forceps of fig. 2. which supports the charcoal, is taken off by unscrewing the screw b; the blow-pipe is also taken off, by loosening the screw n; the hoop fig. 7. is put in its place, where the metallic basin of fig. 19. is put filled with sand; the piece of fig. 8. is set on the other pillar rs, fig. 1. to hold the matrass, fig. 18. upright, or the receiver fig. 20, &c.

In the same manner, the retort, fig. 9. may be put in the sand-bath instead of the matrass, with its receiver fig. 20. which may be supported on a bit of cork or wood, hollowed to its figure, and held by the pliers, instead of the charcoal fig. 2.

But if the operation is to be made in the naked fire,

MINERALOGY.

fire, the neck of the retort, fig. 9. being luted to the receiver, or balloon, fig. 20. may be hanged by a little chain with its ring over the flame, being suspended from the piece of fig. 7. or 8. screwed to either of the pillars as may be most convenient. Otherwise the receiver, fig. 20. may be supported by the round hoop of brass, fig. 8. or 7. screwed at a proper height to the pillar, fig. 1. tying round it some packthread to defend the glass from the contact with the metallic support.

The piece of fig. 6. may be screwed by its collar and screw *ef* to any of the pillars; carrying with it the retort and its receiver, at proper distances, higher or nearer to the lamp according as the flame is more or less violent.

It easily may be conceived, that these implements afford all sorts of conveniences for making any kind of small operations and assays in miniature, provided the operator pays a proper attention to the disposition requisite for each process or operation.

Every glass retort, receiver, matrass, bason, small funnels, &c. are made by the lamp-workers, that blow beads, thermometers, and other small glass instruments.

It is directed that the lamp *k*, fig. 3. be filled with spirit of wine, because it gives no disagreeable smell, and does not produce any fuliginous and disagreeable crust on the vessels as oil does: moreover, the spirit gives a dry flame, without smoke, and stronger than oil; besides the spots and disagreeable consequences this last causes, if split, &c. M. de Morveau adds, that the expence of spirit is quite inconsiderable; and that he performed in eight or ten minutes, with this apparatus, various dissolutions, evaporations, and other processes, which otherwise would have taken more than three hours, with the expence only of two or three halfpence for the spirit of wine, whilst the fuel of charcoal would have cost near ten or eleven pence.

But a very important circumstance is, as Morveau observes likewise, that many philosophers do not apply themselves to chemical operations, for want of opportunity of having a laboratory to perform them: it requiring a proper room, and suitable expences of many large furnaces, retorts, crucibles and numerous other implements, &c. whilst these miniature laboratories may in great measure afford the same advantages; at least to that degree of satisfaction sufficient to ascertain the contents and products of any substance that is subjected to trial: for with this simple apparatus a man of some abilities may, without any embarassment, in a very short time, and with little expence, perform such distillations as require a reverberatory furnace; all sorts of processes, digestions, and evaporations, which require a regular sand heat; he may vary his experiments or trials, and multiply them to a great number of various performances, draw up his conclusions, and reason upon them, without loss of time, without the hinderance of long preparations to work at large. And even when such large works are to be performed, he may observe beforehand various phenomena of some substances, which being known in time, would otherwise impede the processes at large, or make them fail absolutely; and all this without the risk of a considerable loss, and without exposing himself to a great fire, &c.

Of Arrangement.

[*Editor's Note:* The remainder of this article (pp. 69–145) details the characteristics of the known minerals and rocks and has been omitted here.]

DESCRIPTION, AND CHEMICAL EXAMINATION OF AN ORE OF ZINC, FROM NEW JERSEY

Archibald Bruce

RED OXIDE OF ZINC.

External Characters.

Colour, light and dark red, approaching to blood red ruby and aurora red.

Opaque, though generally translucent on the edges.

Fracture foliated—cross fracture slightly conchoidal.

Lustre on fresh fracture, shining: after long exposure to the atmosphere, dull: the surface in time becoming covered with a pearl white crust.

Brittle—being easily pulverised, the powder, brownish yellow, approaching to orange.

Readily scratched by steel.

Specific gravity, 6.22.

Chemical Characters.

Soluble in the mineral acids.

Infusible before the blow-pipe per se. With Sub-Borate of soda, melts into a transparent yellow bead. When exposed to the united flames of oxygene and hydrogene,* it sublimes, attended with a brilliant white light.

When powdered, and with potash exposed to heat, it fuses into an emerald green mass, which on solution, affords to water the same colour. On the addition of a few drops of nitric, sulphuric or muriatic acids, the green coloured fluid is immediately changed to a rose red.

Distinguishing Characters.

Its infusibility distinguishes it from the *red antimoniated sulphuret of silver*, which is fusible before the blow-pipe, giving out white fumes, and a yellow tinge to the charcoal, leaving a globule of reduced silver.

From the *ruby red oxide of copper*, it differs in weight, being nearly twice as heavy, its specific gravity being only

---

* For the application of hydrogene and oxygene gasses, to the purposes of the blow-pipe, we are indebted to Robert Hare, Jun. Esq. Professor of Natural Philosophy in the University of Pennsylvania. For a description of his hydrostatic blow-pipe, we refer the reader to a pamphlet published in 1802, by order of the Chemical Society of Philadelphia; and also to a paper of his, read before the American Philosophical Society, in June, 1803, and published in the sixth volume of their Transactions.

3.9. It is also distinguished by its solution in acids, being colourless; whereas that of the red oxide of copper is of a bright green. Its solubility in the mineral acids is sufficient to distinguish it from the *red oxide of titanium*, which is insoluble. The *red chromate of lead*, before the blow-pipe, melts into a blackish slag. The *red sulphuret of arsenic* is volatilized, giving a blue flame, and a strong smell of garlic.

Locality.

This mineral occurs in several of the iron mines in Sussex county, New-Jersey; as at the Franklin, Stirling, and Rutgers mines, and near Sparta. In some instances, it is imbedded in a sparry lime stone; while in others, it serves as the matrix of several varieties of octaedral oxide of iron, which sometimes occurs crystallized, though more generally in various sized irregular grains.

At Franklin, it also assumes a micaceous form, and is imbedded in a whitish oxide of zinc, which is often, in the same specimen, found adhering to the black oxide of iron.

Having ascertained by previous experiments, that this mineral was principally composed of oxide of zinc, iron and manganese, it was submitted to the following

Chemical Examination.

A. Twenty-five grains of the ore, in fine powder, were dissolved in diluted nitric acid : the solution was colourless.

B. To the solution A, was added oxalic acid, while any precipitation took place. The precipitate (oxalate of zinc) being separated by the filter and dried, weighed 42 grains.

C. The filtered fluid being evaporated to dryness, a dark brown coloured mass remained, which after being ignited, weighed two grains.

D. The brown coloured mass C, being dissolved in diluted muriatic acid, into the solution was dropped a solution of super-tartrate of potash. After standing a few minutes, the solution became turbid, and a precipitation of minute crystalline grains (tartrate of manganese) took place. To the remaining fluid was added Prussiate of potash, which produced a dark blue colour; and after a few minutes a blue powder (Prussiate of iron) was precipitated.

E. The oxalate of zinc, B, (consisting of 23 grains oxide of zinc, and 19 of oxalic acid) was exposed to a low red heat (in a platina crucible) for ten minutes, when the powder was changed to a light yellow colour. On further ignition, for half an hour, the colour became darker, and the powder on being weighed was found to have lost 20 grains, the remaining 22 grains being a sub-oxide of zinc, which according to the experiments of Desormes and Clements, contains in the hundred parts 88 of zinc, and 12 of oxigene.

According to this examination, one hundred parts of the ore contain

Zinc	76
Oxigene	16
Oxides of Manganese and Iron	8
	100

[*Editor's Note:* A brief concluding note on the refining of zinc ores (pp. 99-100) has been omitted here.]

Review of an elementary Treatise on Mineralogy and Geology, being an introduction to the study of these sciences, and designed for the use of pupils; for persons attending lectures on these subjects; and as a companion for travellers in the United States of America—Illustrated by six plates. By PARKER CLEAVELAND, *Professor of Mathematics and Natural Philosophy, and Lecturer on Chemistry and Mineralogy in Bowdoin College, Member of the American Academy, and Corresponding Member of the Linnæan Society of New England.*

Benjamin Silliman

THIS work has been for some time before the public, and it has been more or less the subject of remark in our various journals. It is, however, so appropriate to the leading objects of *this* Journal, that we cannot consider ourselves as performing labours of supererogation while we consider the necessity, plan, and execution of the treatise of Professor Cleaveland.

An extensive cultivation of the physical sciences is peculiar to an advanced state of society, and evinces, in the country where they flourish, a highly improved state of the arts, and a great degree of intelligence in the community. To this state of things we are now fast approximating. The ardent curiosity regarding these subjects, already enkindled in the public mind, the very respectable attainments in science which we have already made, and our rapidly augmenting means of information in books, instruments, collections, and teachers, afford ground for the happiest anticipations.

Those sciences which require no means for their investigation beyond books, teachers, and study—those which demand no physical demonstrations, no instruments of research, no material specimens: we mean those sciences which relate only to the intellectual and moral character of man, were early fostered, and, in a good degree, matured in this country. Hence, in theology, in ethics, in jurisprudence, and in civil policy, our advances were much earlier, and more worthy of respect, than in the sciences relating to material things. In some of these, it is true, we have made very considerable advances, especially in natural philosophy and the mathematics, and their applications to the arts; and this has been true, in some good degree, for very nearly a century. Natural history has been the most tardy in its growth, and no branch of it was, till within a few years, involved in such darkness as mineralogy. Notwithstanding the laudable efforts of a few gentlemen to excite some taste for these subjects, so little had been effected in forming collections, in kindling curiosity, and diffusing information, that only fifteen years since, it was a matter of extreme difficulty to obtain, *among ourselves*, even *the names* of the most common stones and minerals; and one might inquire earnestly, and long, before he could find any one to identify even *quartz*, *feldspar*, or *hornblende*, among the simple minerals; or *granite*, *porphyry*, or *trap*, among the rocks. *We speak from experience*, and well remember with what impatient, but almost despairing curiosity, we eyed the bleak, naked ridges, which impended over the valleys and plains that were the scenes of our youthful excursions. In vain did we doubt whether the glittering spangles of mica, and the still more alluring brilliancy of pyrites, gave assurance of the existence of the precious metals in those substances; or whether the cutting of glass by the garnet, and by quartz, proved that these minerals were the diamond; but if they were not precious metals, and if they were not diamonds, we in vain inquired of our companions, and even of our teachers, what they were.

We do not forget that Dr. Adam Seybert, in Philadelphia; Dr. Samuel L. Mitchill, in New-York; and Dr. Benjamin Waterhouse, in Harvard University, began at an earlier period to enlighten the public on this subject; they began to form collections; Harvard received a select cabinet from France and England; and Mr. Smith, of Philadelphia, (although, returning from Europe fraught with scientific acquisitions, he perished tragically near his native shores,) left his collection to enrich the Museum of the American Philosophical Society.

Still, however, although individuals were enlightened, no serious impression was produced on the public mind; a few lights were indeed held out, but they were lights twinkling in an almost impervious gloom.

The return of the late Benjamin D. Perkins, and of the late Dr. A. Bruce, from Europe, in 1802 and 3, with their collections, then the most complete and beautiful that this country had ever seen; the return of Colonel Gibbs, in 1805, with his extensive and magnificent cabinet; his consequent excursions and researches into our mineralogy; the commencement, about this time, of courses of lectures on mineralogy, in several of our colleges, and of collections by them and by many individuals; the return of Mr. Maclure, in 1807; his Herculean labour in surveying the United States geologically, by personal examination; and the institution of the American Journal of Mineralogy, by Dr. Bruce, in 1810;—these are among the most prominent events, which, in the course of a few years, have totally changed the face of this science in the United States.

During the last ten years, it has been cultivated with great ardour, and with great success: many interesting discoveries in American mineralogy have been made; and this science, with its sister science, Geology, is fast arresting the public attention. In such a state of things, books relating to mineralogy would of course be eagerly sought for.

No work, anterior to Kirwan, could be consulted by the student with much advantage, on account of the wonderful progress, which, within forty or fifty years, has been made in mineralogy. Even Kirwan, who performed a most important service to the science, was become, in some considerable degree, imperfect and obsolete; the German treatises, the fruitful fountains from which the science had flowed over Europe, were not translated; neither were those of the French; and this was the more to be regretted, because they had mellowed down the harshness and enriched the sterility of the German method of description, besides adding many interesting discoveries of their own. It is true we possessed the truly valuable treatise of Professor Jameson, the most complete in our language. But the expense of the work made it unattainable by most of our students, and the undeviating strictness with which the highly respectable author has adhered to the German mode of description, gave it an aspect somewhat repulsive to the minds of novices, who consulted no other book. We are, however, well aware of the value of this work, especially in the improved edition. It must, without doubt, be in the hands of every one who would be master of the science; but it is much better adapted to the purposes of proficients than of beginners.

The mineralogical articles dispersed through Aikin's Dictionary are exceedingly valuable; but, from the high price of the work, they are inaccessible to most persons.

The most recent of the French systems, that by Brongniart, seemed to combine nearly all the requisites that could be desired in an elementary treatise; and a translation of it would probably, ere this, have been given to the American public, had we not been led to expect the work of Professor Cleaveland, which, it was anticipated, would at least possess one important advantage over the work of Brongniart, and every other; it would exhibit, more or less extensively, *American localities*, and give the leading features of our natural mineral associations.

Thus it appears* that the work of Professor Cleaveland was eminently needed; the science, at large, needed it; and to American mineralogists it was nearly indispensable. It appeared too at a very opportune moment. Had it come a few years sooner, in might not have found many readers. Now it is sustained by the prevailing curiosity, and diffused state of information regarding mineralogy; and, in turn, no cause could operate more effectually to cherish this curiosity, and to diffuse this information still more widely, than this book. Professor Cleaveland is therefore entitled to our thanks for undertaking this task; and, in this age of book-making, it is no small negative praise if an author be acquitted of *unnecessarily* adding to the already onerous mass of books.

With respect to the PLAN of this work, Professor Cleaveland has, with good judgment, availed himself of the excellencies of both the German and French schools.

Mr. Werner, of Fribourg, in some sense not only the founder of the modern German school of mineralogy, but almost of the science itself, is entitled to our lasting gratitude for his system of external characters, first published in 1774. In this admirable treatise he has combined precision and copiousness, so that exact ideas are attached to every part of the descriptive language, and every character is meant to be defined.

It is intended that a full description of a mineral upon this plan shall entirely exhaust the subject, and that although many properties may be found in common among different minerals, still every picture shall contain *peculiar* features, not to be found in any other. It would certainly appear, at first view, that this method must be perfect, and leave nothing farther to be desired. It has, however, been found in practice, that the full descriptions of the Wernerian writers are heavy and dry; they are redundant also, from the frequent repetition

---

* The smaller works of Phillips and Aikin were not then published; had they been, they could not have superseded Cleaveland; the same may be said of the respectable work of Professor Kidd, of Oxford University.

of similar properties; and from not giving due prominence to those which are peculiar, and therefore distinctive, they frequently fail to leave a distinct impression of any thing on the mind, and thus, in the midst of what is called by the writers of this school a full *oryctognostic picture*, a student is sometimes absolutely bewildered.

Some of the modern French writers, availing themselves of Mr. Werner's very able delineation of the external characters of minerals, have selected such as are most important, most striking, distinctive, and interesting; and drawing a spirited and bold sketch, have left the minuter parts untouched: such a picture, although less perfect, often presents a stronger likeness, and more effectually arrests the attention.

This is the method of description which has been, as we think, *happily* adopted, to a great extent by Mr. Cleaveland.

Mr. Werner, availing himself of the similarities in the external appearance of minerals, has (excepting the metals) *arranged* them also upon this plan, without regard to their constitution; that is, *to their real nature*, or, at least, making this wholly subservient to the other: this has caused him, in some instances, to bring together things which are totally unlike in their nature, and, in other instances, to separate those which were entirely similar. Whatever may be said in favour of such a course, considered as a provisional one, while chemical analysis was in its infancy, the mind can never rest satisfied with any arrangement which contradicts the real nature of things; in a word, the composition of minerals is the only correct foundation for their classification. This classification has been adopted by several of the ablest modern French writers.

"It is believed," (says Professor Cleaveland, Preface, p. 7.) "that the more valuable parts of the two systems may be incorporated, or, in other words, that the peculiar descriptive language of the one may, in a certain degree, be united to the accurate and scientific arrangement of the other.

"This union of descriptive language and scientific arrangement has been effected with good success, by BRONGNIART, in his System of Mineralogy—an elementary work, which seems better adapted both to interest and instruct, than any which has hitherto appeared. The author of this volume has, therefore, adopted the *general* plan of Brongniart, the more important parts of whose work are, of course, incorporated with this."

A happier model could not, in our opinion, be chosen; and we conceive that Professor Cleaveland is perfectly consistent, and perfectly perspicuous, when, adopting the chemical composition of minerals as the only proper foundation of arrangement, and, of course, rejecting the principle of Mr. Werner, which arranges them upon their external properties, he still adopts his *descriptive* language as far as it answers his purpose. For to elect a principle of arrangement, and to classify all the members of a system so as to give each its appropriate place, is obviously quite a different thing from describing each member, after its place in a system is ascertained. In doing the latter, characters may be drawn from any source which affords them.

In his " Introduction to the Study of Mineralogy," the author has given a view at once copious, condensed, and perspicuous, of all that is necessary to be learned previously to the study of particular minerals. He begins with definitions and general principles, which are laid down with clearness.

By way of engaging the attention to the study of this department of nature, he remarks:

" From a superficial view of minerals in their natural depositories, at or near the surface of the earth, it would hardly be expected that they could constitute the object of a distinct branch of science. Nothing appears farther removed from the influence of established principles and regular arrangement, than the mineral kingdom when observed in a cursory manner. But a closer inspection and more comprehensive view of the subject will convince us, that this portion of the works of nature is by no means destitute of the impress of the Deity. Indications of the same wisdom, power, and benevolence, which appear in the animal and vegetable kingdoms, are also clearly discernible in the mineral."

" It may also be remarked," continues the author, " that several arts and manufactures depend on mineralogy for their existence; and that improvements and discoveries in the latter cannot fail of extending their beneficial effects to the aforementioned employments. In fine, the study of mineralogy, whether it be viewed as tending to increase individual wealth, to improve and multiply arts and manufactures, and thus promote the public good; or as affording a pleasant subject for scientific research, recommends itself to the attention of the citizen and scholar."

This introductory view of the importance and interest of the science cannot be charged with the fault of exaggeration, since it is most evident that neither civilization, refinement in arts, nor comfort, can exist where the properties of mineral substances are but imperfectly understood.

As regards this country, the argument admits of much amplification. The more our mineral treasures are explored, the more abundantly do they repay the research; and we trust that the period is not far distant, when we shall no longer ignorantly tread under our feet minerals of great curiosity and value, and import from other countries, at a great expense, what we, in many instances, possess abundantly at home.*

But to return to the plan of the author's work. Few persons, unacquainted with the science of mineralogy, would suspect that mere brute matter could exhibit many strong marks, capable of discrimination.

It may, however, be confidently affirmed, that there is no mineral which, if carefully studied, may not be distinguished by characters sufficiently decisive from every other mineral; an account of these characters ought, therefore, to precede every system of mineralogy. Professor Cleaveland has, with entire propriety, included them under the heads of crystal-

---

* A vast region in the interior of New-York and Pennsylvania is now fertilized by inexhaustible beds of sulphat of lime, (plaster of Paris,) which, till a very few years since, were not even known to exist.

Near New-Haven immense beds of green marble were discovered in 1811, during a mineralogical excursion: this beautiful material, closely resembling the *verd antique*, is now, on the spot, wrought into tables, fireplaces, and many other ornamental forms; and although the farmers had made fences of it for 150 years, no one suspected what it was till the study of mineralogy, in Yale College, brought it to light.

iography, physical and external characters, and chemical characters.

He has given a clear view of the Abbé Haüy's curious discoveries regarding the six primitive figures or solids which form the bases of all crystals—the three integrant particles or molecules which constitute the primitive forms, and of the theory by which it is shown how the immensely numerous and diversified secondary or actual forms arise out of these few elementary figures.

This is certainly one of the most singular and acute discoveries of our age. It is true, there is a difference of opinion among mineralogists as to the practical use of crystallography in the discrimination of minerals. Some dwell upon it with excessive minuteness, and others seem restless and impatient of its details. The truth seems to be, that those who understand it, derive from it (wherever it is applicable) the most satisfactory aid; and it requires only a moderate knowledge of geometry to understand its principal outlines. On the other hand, it is no doubt possible, in most instances, to dispense with its aid, and to discriminate minerals by their other properties.

Of the external and physical characters of Mr. Werner, Mr. Cleaveland has given a clear account, combining into the same view the fine discriminations of the French authors, particularly regarding refraction, phosphorescence, specific gravity, electricity, chatoyement, and magnetism. The same may be said of the chemical characters. We do not know a more satisfactory and able view of the characters of minerals than Professor Cleaveland has exhibited.

We would however ask, whether, in enumerating the kinds of lustre, the term *adamantine* should not be explained, as it is not understood by people in general, while the terms denoting the other kinds are *generally* intelligible; whether in the enumeration of imitative forms, *lenticular* and *acicular* should not rather be referred to the laws of crystallization; whether *reniform* and *mamillary* are synonymous; whether *sandstone*, as being a mere aggregate of *fragments*, is a good instance of the *granular* fracture; whether in its natural state (at least the common ore of nickel) is *ever* magnetic, till *purified*, and whether cobalt is *ever* magnetic unless *impure*.

Professor Cleaveland's remarks on *fracture* are uncommonly discriminating and instructive, and would lead a learner to a just comprehension of this important point in the characters of minerals.

The section relating to the *chemical characters* is concise, and professedly proceeds upon the principle of selection. It might perhaps have been, to some extent, advantageously enlarged; although, it is true, the author refers us to the particular minerals for individual instances; still it might have been well to have illustrated the general principles by a few well-chosen instances, *e. g.* how, by the blowpipe, *galena* is distinguished from *sulphuret of antimony; carbonat of lead* from *sulphat of barytes*, or *carbonat of lime; garnet* from *titanium; plaster of Paris* from *soapstone*, &c.; and, among trials in the moist way, how by nitric acid and ammonia, *iron pyrites* is distinguished from *copper pyrites;* and how, by acids, *sulphat of lime* is known from *carbonat of lime*. As the acids are used principally for trials on the effervescence of carbonats, most of which form with sulphuric acid, insoluble compounds, we should doubt whether sulphuric acid is so advantageously employed as the nitric or muriatic, in such cases, on account of the clogging of the effervescence by the thick magena, produced by a recently precipitated and insoluble sulphat.

According to our experience, the nitric or muriatic acid, diluted with two or three parts of water, is most eligible.

With respect to the blowpipe: it is *a convenience* to have a mouth-piece of wood, or ivory, joined to a tube of metal, as Mr. Cleaveland recommends; and some authors direct to have the tube attached to a hollow ball, for the sake of condensing the moisture of the breath; but every thing which adds to the expense and complication of the instrument will tend to discourage its use; we have never found any difficulty in performing every important experiment with the common goldsmith's brass blowpipe; and are confident, that, after the learner has acquired the art, or *knack*, of propelling a continued stream of air from his mouth, by means of the muscles of the lips and cheeks, while his respiration proceeds without embarrassment through the nostrils, he will need no other instrument than the common blowpipe. Indeed it is a truly admirable instrument, instantly giving us the effect of very powerful furnaces, the heat being entirely under command, the subject of operation and all the changes in full view, and the expense and bulk of the instrument being such that every one may possess it, and carry it about his person.

The chapter on the principles of arrangement is worthy of all praise. This difficult subject is here discussed with such clearness, comprehensiveness, and candour, as prove the author to be completely master of his subject; and we are persuaded, that, on this topic, no author can be studied with more advantage. We forbear to extract, because the whole should be attentively perused in connexion, and scarcely admits of abridgement. We entirely agree with Professor Cleaveland, as we have already said, that the chemical composition of minerals is the only just foundation of their arrangement; that next in importance is the crystalline structure, including a knowledge of the primitive form, and integrant molecule; and last and least important, *in fixing the arrangement*, are the external characters: these last should be only provisionally employed, where the two first are not ascertained, or the second is not applicable. When the arrangement is once made, we *may*, however, and we commonly *shall*, in *describing* minerals, pursue precisely the reverse order; the external characters will usually be mentioned first, the crystalline characters next, and the chemical last of all. In description, the external characters are often the most valuable; if judiciously selected and arranged, they will always prove of the most essential service, and can rarely be entirely dispensed with.

With regard to the NOMENCLATURE of minerals, we feelingly unite with Professor Cleaveland in deploring the oppressive redundancy of synonymes. Few minerals have only one name, and usually they have several. With Count Bournon we agree, that the discoverer of a mineral has the exclusive right of naming it, and that the name once given should not be changed without the most cogent reasons. What then shall we say of the ABBE HAÜY, of whom, whether we speak of his genius, his learning, his acuteness, his discoveries, his candour, and love of truth, or his universally amiable and venerable character, we can never think without sentiments of the highest respect and admiration? More than any modern writer he has added to the list of synonymes, often exchanging a very good name, derived perhaps from the locality or discoverer of a mineral, for one professedly significant, but connected with its subject by a chain of thought so slight, that considerable knowledge of Greek etymology, and still more explanation, is necessary to comprehend the connexion; and thus, after all, it amounts, with respect to most readers, only to the exchange of one arbitrary name for another. What advantage, for instance, has *grammatite*, alluding to a line often obscure, and still oftener wholly invisible, over the good old name *tremolite*, which always reminds us of an interesting locality; how is *pyroxene* better than *augite*, *amphibole* than *hornblende*, *amphigene* than *leucite*, or *disthene* than *sappar*. Some of the Abbé Haüy's names are, however, very happily chosen, especially where new discriminations were to be established, or errors corrected, or even a redundant crop of synonymes to be superseded by a better name. *Epidote* is an instance of the latter, and the new divisions of the old *zeolite family* into four species, *mesotype*, *stilbite*, *analcime*, and *chabasie*, afford a happy instance of the former. It were much to be wished, that by the common consent of mineralogists, one nomenclature should be universally adopted: for its uniformity is of much more importance than its nature.

In expressing our approbation of the principles of arrangement adopted by Professor Cleaveland, we have of course espoused those of his TABULAR VIEW, which is perhaps as nearly as the state of science will admit, erected upon a chemical basis, like that of Brongniart, to which it bears a close resemblance. Some of the subordinate parts, we could have wished had been arranged in a manner somewhat different. In the genus lime, it appears to us better to describe the species carbonat first; because, being very abundant, and its characters clear, it forms a convenient point of departure and standard of comparison, in describing the other species which have lime for their basis, and some of which are comparatively rare. The same remark we would make upon quartz, and its concomitant, pure silicious stones. There appears to us a high advantage in making these minerals clearly known first, before we proceed to those which are much more rare, and especially which are much harder, and possess the characters of gems. For example, if a learner has become acquainted with quartz, chalcedony, flint, opal, chrysoprase, and jasper, he will much more easily comprehend the superior hardness, &c. and different composition of topaz, sapphire, spinelleruby, chrysoberyl, and zircon, which we should much prefer to see occupying a later, than the first place in a tabular arrangement; and, although topaz, by containing fluoric acid, appears to be in some measure assimilated to saline minerals, it is in its characters so very diverse from the earthy salts, that we have fair reason to conclude that the fluoric acid does not stamp the character; and, as it bears so close a resemblance to the ruby and sapphire, which evidently derive their principal characters from the argillaceous earth, we perhaps ought to infer that this (the topaz,) does so too. Indeed Professor Cleaveland has sufficiently implied his own opinion, by giving these minerals a juxtaposition in his table, although the same reasons which induced the placing of the topaz next to the earthy salts, could not have justified the placing of the sapphire there. On these points we are not, however, strenuous; they are of more importance if the work be used as a text-book for lectures, than as a private companion. With respect to the *completeness* of Professor Cleaveland's tabular view, we have carefully compared it with the third edition of Jameson's mineralogy; and although a few new species, or sub-species, and varieties have been added in this last edition, they are in general of so little importance, that Professor Cleaveland's work cannot be considered as materially deficient; and the few cases in which it is so, are much more than made up by his entirely new and instructive views of American mineralogy, to which no parallel is to be found in any other book, and which give it peculiar interest to the American, and even to the European, reader.

In another edition, (which we cannot doubt will speedily be called for,) he will of course add whatever is omitted in this, and we should be gratified to see a good article on the subject of the aerolites or stones which have fallen from the atmosphere. This subject is one, in our view, of high interest; and although *in strictness* it may not claim a place in a tabular view of minerals, (we must confess, however, that we see no important obstacle to its being treated of under the head of native iron,) there can be no objection to its being placed in an appendix. The fall of stones from the atmosphere is the most curious and mysterious fact in natural history.

It may seem perhaps too trivial to remark, that the annexation of numbers, referring to the pages, would be a serious addition to the utility of the tabular view. Very few inadvertencies have been observed—the following may be mentioned: *Amenia*, in the State of New-York, is printed (by a typographical error we presume) Armenia; and *Menechan*, where the menechanite is found, is mentioned as occurring in Scotland, but it is in Cornwall.

Authors seem agreed that the black-lead ore is an altered carbonat, but they seem not to have been so well agreed as to the nature of the blue-lead ore. In the cabinet of Colonel Gibbs, there are specimens which appear satisfactorily to illustrate both these subjects. The black-lead is by the blowpipe alone reducible to metallic lead; there is one specimen in the cabinet referred to, which is blackened on what appears to have been the under side, and seemingly by the contact of sulphuretted hydrogen gas; that which was probably the upper part remains unaltered, and is beautiful white carbonat of lead; this appearance is the more striking, because the piece is large and full of interstices, by which the gas appears to have passed through. The blue ore is in large six-sided prisms of a dark blue or almost black colour; where the prisms are broken across, they present an unequal appearance; sometimes they are *invested*; and sometimes slightly, and at other times deeply, *penetrated* by sulphuret of lead, having the usual brilliant foliated fracture. The part which looks like sulphuret of lead is easily reducible by the blowpipe, but not the whole crystal, as authors appear to imply; for if that part of the crystal which does not present the appearance of galena is heated by the blowpipe flame, it is not reduced, but congeals into the garnet dodecahedron, with its colour unaltered: these crystals are therefore phosphat of lead, and they appear to be either an original mixture of phosphat and sulphuret of lead, or the phosphat has somehow in part given up its phosphoric acid, and assumed in its stead sulphur, perhaps from the decomposition of sulphuretted hydrogen.

Professor Cleaveland will, of course, add new localities, even foreign ones, where they are interesting, and domestic ones, where they are well authenticated. Among the former, we trust he will mention the lake of sulphuric acid contained in the crater of Mount Idienne, in the Province of Bagnia Vangni, in the eastern part of Java, and also the river of sulphuric acid which flows from it and kills animals, scorches vegetation, and corrodes the stones.* Among American localities, we beg leave to mention violet fluor spar, abundant and very handsome, near Shawnee Town, on the Ohio, in the Illinois Territory, and galena, of which this fluor is the gangue;—sulphat of magnesia, perfectly crystallized, in masses composed of delicate white prisms, in a cave in the Indiana Territory, not very remote from Louisville, in Kentucky; it is said to be so abundant that the inhabitants carry it away by the wagon load;—pulverulent carbonat of magnesia, apparently pure, found by Mr. Pierce at Hoboken, in serpentine, where the hydrate of magnesia was found;—chabasie, agates, chalcedony, amethyst, and analcime, at Deerfield, by Mr. E. Hitchcock;—agates in abundance at East-Haven, near New-Haven, in secondary greenstone, like the above-named minerals at Deerfield;—saline springs, covered with petroleum, and emitting large volumes of inflammable gases, numerous in New-Connecticut, south of Lake Erie;—magnetical pyrites, abundant in the bismuth vein, at Trumbull, Connecticut;—very

* See Tilloch's Phil. Mag. Vol. XLII. p. 182.

brilliant fine-grained micaceous iron, in large masses near Bellows' Falls; yellow foliated blende, in Berlin, Connecticut, and near Hamilton College—the latter discovered by Professor Noyes; it is in veins in compact limestone;—red oxid of titanium, often geniculated, at Leyden, in Massachusetts, discovered by Mr. E. Hitchcock;—red oxid of titanium, in very large crystals and geniculated, imbedded in micaceous schistus, at Oxford, 20 miles north from New-Haven;—silicious petrifactions of wood, abundant in the island of Antigua, recently brought by Mr. Pelatiah Perit, of New-York;—sulphuret of molybdena, at Pettipaug, and at East-Haddam, Connecticut;—prehnite abundant and beautiful, in secondary greenstone, at Woodbury, 24 miles north of New-Haven, discovered by Mr. Elijah Baldwin;—black oxid of manganese, in great abundance, and of an excellent quality, near Bennington, Vermont, and plumose mica, in a very fine graphic granite, in a hill two miles north of Watertown, Connecticut.

The introduction to the STUDY OF GEOLOGY, deserves a more extended series of remarks than it would now be proper to make, after so full a consideration of the previous parts of the work.

Professor Jameson's elaborate exposition of the Wernerian system, is too full, and too much devoted to a particular system, for beginners: the sketches of geology contained in the systems of Chemistry by Murray and Thomson, and in Phillips's mineralogy, are two limited, although useful: the excellent account of the Wernerian system, contained in an Appendix to Brochant's Mineralogy, has, we believe, never been translated; and we need not say that Professor Playfair's illustrations of the Huttonian Theory, De Luc's Geology, and Cuvier's Geology, are not well adapted to the purposes of a beginner; neither is Delametherie's, nor has it been translated. An introduction to geology was, therefore, hardly less needed than one to mineralogy. Professor Cleaveland has performed this difficult duty with great ability, and has brought this interesting branch of science fairly within the reach of our students.

Although adhering substantially to the Wernerian arrangement of rocks, he has, so to speak, blended Werner's three classes of primitive, transition, and secondary rocks, into one class; and where the same rock occurs in all the three classes, or in two of them, he mentions it in giving the history of the particular rock. This method simplifies the subject very much to the apprehensions of a learner. A rigid Wernerian would probably revolt at it, but the distinctions of Mr. Werner may still be pointed out, and, we should think, ought to be, at least by all teachers.

In Mr. Cleaveland's account of the trap rocks, we should almost imagine that some typographical error had crept into the following paragraph:

"But in modern geological inquiries, the word trap is usually employed to designate a *simple mineral*, composed of hornblende nearly or quite pure, and also those aggregates in which *hornblende* predominates. Hence, the *presence* of hornblende, as a predominating ingredient, characterizes those MINERALS to which most geologists apply the name *trap*."

Now, it is not accordant with our apprehensions that trap is ever at the present time employed to designate a *simple mineral*, nor has Professor Cleaveland himself used it in his tabular view, or in his description of simple minerals. In our view, it is the *classical* word of modern geology, to designate that description of rocks in which hornblende predominates, and perhaps a few others of minor importance usally associated with them. It is true, a rock composed of pure hornblende may be called trap, but it is not true, *vice versa*, that this rock, considered in its character of a simple mineral, is called trap. If our views are correct, the section which is headed *trap* or *hornblende*, should be *trap* or *hornblende rocks*, and greenstone should come in as a subdivision, and not form a distinct section. With these alterations, and with the substitution of rock in the *first*, and rocks in the *second* instance, in the paragraph above quoted, instead of *mineral* and *minerals*, we apprehend the view of this family of rocks would be much more clear, and a degree of confusion which learners now experience from the paragraph, would be prevented. If we are wrong, we are sure Professor Cleaveland will pardon us; if right, his candour will readily admit the correction.

As to the manner in which the work of Professor Cleaveland is executed, the remarks which we already made, have in a good degree anticipated this head.

We cannot, however, dismiss the subject without adding that, in our opinion, this work does honour to our country, and will greatly promote the knowledge of mineralogy and geology, besides aiding in the great work of disseminating a taste for science generally. Our views of the plan we have already detailed. The manner of execution is masterly. Discrimination, perspicuity, judicious selection of characters and facts, and a style chaste, manly, and comprehensive, are among the characteristics of Professor Cleaveland's performance. It has brought within the reach of the American student the excellencies of Kirwan, Jameson, Haüy, Brochant, Brongniart, and Werner; and we are not ashamed to have this work compared with their productions. In our opinion Professor Cleaveland's work ought to be introduced into all our schools of mineralogy, and to be the travelling companion of every American mineralogist.

We trust that all cultivators of mineralogy and geology in this country, will willingly aid Professor Cleaveland in enlarging his list of American localities for a second edition; and we hope that he will repay them, at a future day, by giving us a distinct treatise on geology, with as particular a delineation as possible of the geological relations of the great North American formations. Mr. Maclure has, with great ability, sketched the outline; but much labour is still needed in filling up the detail.

Part V

PHYSICS OF THE EARTH

Editor's Comments
on Papers 13 Through 16

13   FRANKLIN
      Conjectures Concerning the Formation of the Earth

14   FRANKLIN
      Queries and Conjectures Relating to Magnetism, and the Theory of the Earth

15   COOPER
      Geology

16   BOWDITCH
      Excerpt from *On the Calculation of the Oblateness of the Earth, by Means of the Observed Lengths of a Pendulum in Different Latitudes, According to the Method Given by La Place in the Second Volume of His "Mécanique Céleste," with Remarks on Other Parts of the Same Work, Relating to the Figure of the Earth*

   The origin of the earth and the nature of its interior are matters for speculation even today. In spite of the lack of data and the uncertain theoretical basis, several naturalists prior to 1820 were willing to propose "theories of the earth." The philosophical spirit of the European Enlightenment naturally led to these queries on the origin and present state of the earth. While the Biblical "seven days" may have satisfied some clergymen, eighteenth-century natural philosophers searched for an earth history more in harmony with known physical laws.
   Three European theories on the formation of the earth and its celestial neighbors were familiar to early Americans and may have influenced their writings. Cecil Schneer (1969) cited René Descartes' *Principia Philosophiae* of 1644 as "the first attempt at a secular account of the earth since classical times." Descartes proposed that planets form from vortices or whorls of matter, which collapse in the presence of gravitational attraction. The earth was merely a small, cooled star in the Cartesian theory, which became a foundation for several subsequent speculations.

Later in the seventeenth century the German mathematician and philosopher Gottfried Wilhelm von Leibnitz characterized the earth as an originally fiery body, which cooled from the outside to leave a hard crust. Condensing vapors formed oceans on the cooling crust, and these waters received the first sediments. Nearly a century later, in 1779, there appeared the *Epoques de la Nature* of George Leclerc, Marquis de Buffon, a naturalist noted earlier in this volume for his opinions on the size of American mammals (Paper 4). Buffon's essay represented "the most fully developed of those theories that began with the earth as a cooling star" (Schneer 1969). His theory was in many respects similar to that of Leibnitz, but Buffon supported his ideas with a detailed quantitative explanation of the formative process. The similar hypotheses of Descartes, Leibnitz, and Buffon formed the basis for subsequent theories of the earth's origin by both European and American scholars.

The earliest American ideas on the internal structure of the earth were contained in several accounts and theories of earthquakes (Papers 1 and 2). These writings, however, only treated the crust of the globe, and made no mention of the earth's interior. More specific reference to the earth's origin is found in the correspondence of Dr. Benjamin Franklin, as published in the *Transactions of the American Philosophical Society* for 1793. As representative to Britain from 1757 to 1775 and later as United States ambassador to France from 1777 to 1785, Franklin had a unique opportunity to discuss natural philosophy with the greatest minds of Europe. The theories of Buffon had been recently published when Franklin wrote letters to the Abbé Soulavie of France (Paper 13) and Governor James Bowdoin of Massachusetts (Paper 14) regarding the formation of the earth and other matters. As a scientist, Franklin is best remembered for experiments on the nature of electricity, but these two letters demonstrate how his active mind touched on many other questions as well.

In his first letter of 1782, Franklin noted that such surface changes as shifting shorelines suggest that the earth's interior is not solid, but rather a dense liquid. This liquid's density, thought Franklin, was caused in part by high pressure and in part by its high iron content, which was also responsible for the earth's magnetic field. The American scientist was quick to assert that conclusions based solely on observations are far more satisfactory than speculations, and that "superior beings smile at our theories, and at our presumption in making them." Nonetheless, Franklin permitted himself "to wander a little in the wilds of fancy."

Franklin's letter to Bowdoin of several years later is somewhat more restrained, being a series of queries rather than an exposition. The questions are admittedly rhetorical, and suggest the author's prejudices. In addition to the ideas proposed in the first letter, Franklin noted that changes in the position of the earth's poles might follow shifts in the dense magnetic fluids. These polar variations, then, might explain occurrences of tropical fossil organisms in northerly climates.

Perhaps the only other early American attempt at a description of the earth's interior was Thomas Cooper's "Outlines of geology" (Paper 15), published as a preface to an excerpt by Robert Jameson on the relative ages of different types of metallic ores. Cooper (1759-1839) was a prolific writer but is best remembered as editor and principal contributor to the *Emporium of Arts and Sciences*, a Philadelphia-based periodical. Cooper's essays usually emphasized the practical applications of chemistry to such useful arts as printing, dyeing, and the processing of ores.

Cooper's "Outlines of geology" began with a clear statement on the differences between Wernerian and Huttonian theories of rock formation. The essay then shifted abruptly to speculations on the nature of the earth's core. The mean specific gravity of the earth was known to be greater than 5 on the basis of experiments that measured the size and mass of the globe. Cooper concluded that the nucleus must contain metal, "probably iron and nickel because these are the only magnetic metals." In addition, he suspected that "some connection exists between the composition of meteorolites, all of which contain iron and nickel," and the chemistry of the earth's interior. In spite of the high density of the earth, Cooper believed in a subterranean network of cavities which he related to earthquakes, as had previous workers (Part 1). The remainder of "Outlines of geology" was devoted to a summary of the Wernerian view of the formation of crustal rocks, based on Jameson's exposition, which will be discussed in the next part.

Theories on the origin and nature of the earth's interior consisted primarily of speculation. Early studies on the shape or "figure" of the earth, on the other hand, were rigorously based on quantitative observations. It was well known in the eighteenth century that the earth was not a perfect sphere but more closely approximated an oblate spheroid. The critical parameter in describing this ellipsoid was the ratio of major-to-minor axes. This problem was tackled by several prominent early American mathematicians, including Robert Adrain (1818), Joseph Clay (1802), Jared Mansfield (1810), and Nathaniel Bowditch (Paper 16).

Nathaniel Bowditch (1773-1838) was a prominent American mathematician and astronomer, but his purely scientific accomplishments have been eclipsed by the fame of the *New American Practical Navigator*, which went through 29 editions prior to 1860. Bowditch's *Practical Navigator* was the bible of ship navigation for North American seamen at a time when American shipping was a thriving industry. Most of Bowditch's original scientific contributions, including his work on the figure of the earth, appear in the *Memoirs* of the Boston-based

American Academy of Arts and Sciences. In addition to his practical and scientific writings, Bowditch prepared an English translation of Laplace's epic *Méchanique Céleste,* on which he based his essay on determining the figure of the earth.

Bowditch commenced by describing the four methods then in use for calculating the earth's deviation from a sphere. All four depended on small differences between large numbers, but the American concluded that the best system was that of Laplace, which was based on "the variation in the lengths of pendulums vibrating in a second of time in different latitudes." The presentation of Laplace's method in the standard English source, Rees' *Encyclopaedia* was, however, in error, and Bowditch presented a revised version of the French mathematician's equations. Agreement between the various methods was significantly improved after these corrections, and Bowditch's value of 1 part in 315 ± 10 is consistent with the modern figure.

REFERENCES

Adrain, R. 1818. Investigation of the Figure of the Earth, and of the Gravity in Different Latitudes. *Am. Philos. Soc. Trans.* n.s. 1:119-135.

Clay, J. 1802. Observations on the Figure of the Earth. *Am. Philos. Soc. Trans.* 5: 312-319.

Mansfield, Jared. 1810. On the Figure of the Earth. *Connecticut Acad. Arts and Sci. Mem.* 1:111-118.

Schneer, C. J., ed., 1969. *Toward a History of Geology.* Cambridge, Mass.: MIT Press.

BIBLIOGRAPHY

Anonymous. 1796. On the Increase of Continents. *Am. Mag. Gen. Repos.* 1:44 (i.e., 34)-38.

Biot, J. B. 1819. Notice of the Operations Undertaken to Determine the Figure of the Earth. *Analectic Mag.* 13:26-41.

Cavendish, H. 1799. Experiments to Determine the Density of the Earth. *Med. Repos.* 2:448-450.

Conjectures concerning the formation of the Earth, &c. in a letter from Dr. B. Franklin, to the Abbé Soulavie.

Passey, September 22, 1782.

SIR,

Read Nov. 21, 1788.

I RETURN the papers with some corrections. I did not find coal mines under the Calcareous rock in Derby Shire. I only remarked that at the lowest part of that rocky mountain which was in sight, there were oyster shells mixed in the stone; and part of the high county of Derby being probably as much above the level of the sea, as the coal mines of Whitehaven were below it, seemed a proof that there had been a great bouleversement in the surface of that Island, some part of it having been depressed under the sea, and other parts which had been under it being raised above it. Such changes in the superficial parts of the globe seemed to me unlikely to happen if the earth were solid to the centre. I therefore imagined that the internal part might be a fluid more dense, and of greater specific gravity than any of the solids we are acquainted

quainted with; which therefore might swim in or upon that fluid. Thus the surface of the globe would be a shell, capable of being broken and disordered by any violent movements of the fluid on which it rested. And as air has been compressed by art so as to be twice as dense as water, in which case if such air and water could be contained in a strong glass vessel, the air would be seen to take the lowest place, and the water to float above and upon it; and as we know not yet the degree of density to which air may be compressed; and M. Amontons calculated, that its density increasing as it approached the centre in the same proportion as above the surface, it would at the depth of——leagues be heavier than gold, possibly the dense fluid occupying the internal parts of the globe might be air compressed. And as the force of expansion in dense air when heated is in proportion to its density; this central air might afford another agent to move the surface, as well as be of use in keeping alive the subterraneous fires: Though as you observe, the sudden rarefaction of water coming into contact with those fires, may also be an agent sufficiently strong for that purpose, when acting between the incumbent earth and the fluid on which it rests.

If one might indulge imagination in supposing how such a globe was formed, I should conceive, that all the elements in separate particles being originally mixed in confusion and occupying a great space, they would as soon as the almighty fiat ordained gravity or the mutual attraction of certain parts, and the mutual repulsion of other parts to exist, all move towards their common centre: That the air being a fluid whose parts repel each other, though drawn to the common centre by their gravity, would be densest towards the centre, and rarer as more remote; consequently all matters lighter than the central part of that air and immersed in it, would recede from the

centre

centre and rife till they arrived at that region of the air which was of the fame fpecific gravity with themfelves, where they would reft; while other matter, mixed with the lighter air would defcend, and the two meeting would form the fhell of the firft earth, leaving the upper atmofphere nearly clear. The original movement of the parts towards their common centre, would naturally form a whirl there; which would continue in the turning of the new formed globe upon its axis, and the greateft diameter of the fhell would be in its equator. If by any accident afterwards the axis fhould be changed, the denfe internal fluid by altering its form muft burft the fhell and throw all its fubftance into the confufion in which we find it.

I will not trouble you at prefent with my fancies concerning the manner of forming the reft of our fyftem. Superior beings fmile at our theories, and at our prefumption in making them. I will juft mention that your obfervation of the ferruginous nature of the lava which is thrown out from the depths of our valcanos, gave me great pleafure. It has long been a fuppofition of mine that the iron contained in the fubftance of this globe, has made it capable of becoming as it is a great magnet. That the fluid of magnetifm exifts perhaps in all fpace; fo that there is a magnetical North and South of the univerfe as well as of this globe, and that if it were poffible for a man to fly from ftar to ftar, he might govern his courfe by the compafs. That it was by the power of this general magnetifm this globe became a particular magnet. In foft or hot iron the fluid of magnetifm is naturally diffufed equally; when within the influence of a magnet, it is drawn to one end of the iron, made denfer there, and rarer at the other, while the iron continues foft or hot, it is only a temporary magnet: If it cools or grows hard in that fituation, it becomes a permanent one, the magnetic fluid not eafily refuming its equilibrium. Perhaps it may be owing to the

A 2   permanent

permanent magnetism of this globe, which it had not at first, that its axis is at present kept parallel to itself, and not liable to the changes it formerly suffered, which occasioned the rupture of its shell, the submersions and emersions of its lands and the confusion of its seasons. The present polar and equatorial diameters differing from each other near ten leagues; it is easy to conceive in case some power should shift the axis gradually, and place it in the present equator, and make the new equator pass through the present poles, what a sinking of the water would happen in the present equatorial regions, and what a rising in the present polar regions; so that vast tracts would be discovered that now are under water, and others covered that now are dry, the water rising and sinking in the different extremes near five leagues.—Such an operation as this, possibly, occasioned much of Europe, and among the rest, this mountain of Passy, on which I live, and which is composed of lime stone, rock and sea shells, to be abandoned by the sea, and to change its ancient climate, which seems to have been a hot one. The globe being now become a permanent magnet, we are perhaps safe from any future change of its axis. But we are still subject to the accidents on the surface which are occasioned by a wave in the internal ponderous fluid; and such a wave is producible by the sudden violent explosion you mention, happening from the junction of water and fire under the earth, which not only lifts the incumbent earth that is over the explosion, but impressing with the same force the fluid under it, creates a wave that may run a thousand leagues lifting and thereby shaking successively all the countries under which it passes. I know not whether I have expressed myself so clearly, as not to get out of your sight in these reveries. If they occasion any new enquiries and produce a better hypothesis, they will not be quite useless. You see I have given a loose to imagination; but I approve much more

your

your method of philosophizing, which proceeds upon actual observation, makes a collection of facts, and concludes no farther than those facts will warrant. In my present circumstances, that mode of studying the nature of this globe is out of my power, and therefore I have permitted myself to wander a little in the wilds of fancy. With greate steem I have the honour to be, &c.

P. S. I have heard that chemists can by their art decompose stone and wood, extracting a considerable quantity of water from the one, and air from the other. It seems natural to conclude from this, that water and air were ingredients in their original composition. For men cannot make new matter of any kind. In the same manner may we not suppose, that when we consume combustibles of all kinds, and produce heat or light, we do not create that heat or light; but only decompose a substance which received it originally as a part of its composition? Heat may thus be considered as originally in a fluid state, but, attracted by organized bodies in their growth, becomes a part of the solid. Besides this, I can conceive that in the first assemblage of the particles of which this earth is composed each brought its portion of the loose heat that had been connected with it, and the whole when pressed together produced the internal fire which still subsists.

QUERIES *and* CONJECTURES *relating to Magnetism, and the Theory of the Earth, in a Letter from* Dr. B. FRANKLIN, *to* Mr. BODOIN,

DEAR SIR,

Read Jan. 15, 1790.

I RECEIVED your favours by Meffrs. Gore, Hilliard and Lee, with whofe converfation I was much pleafed, and wifhed for more of it; but their ftay with us was too fhort. Whenever you recommend any of your friends to me, you oblige me.

I want to know whether your Philofophical Society received the fecond volume of our Tranfactions. I fent it, but never heard of its arriving. If it mifcarried, I will fend another. Has your Society among its books the French Work *fur les Arts & les Metiers?* It is voluminous, well executed, and may be ufeful in our country. I have bequeathed it them in my will; but if they have it already, I will fubftitute fomething elfe.

Our

Our ancient correspondence used to have something philosophical in it. As you are now more free from public cares, and I expect to be so in a few months, why may we not resume that kind of correspondence? Our much regretted friend Winthrop once made me the compliment, that I was good at starting game for philosophers, let me try if I can start a little for you.

Has the question, how came the earth by its magnetism, ever been considered?

Is it likely that *iron ore* immediately existed when this globe was first formed; or may it not rather be supposed a gradual production of time?

If the earth is at present magnetical, in virtue of the masses of iron ore contained in it, might not some ages pass before it had magnetic polarity?

Since iron ore may exist without that polarity, and by being placed in certain circumstances may obtain it, from an external cause, is it not possible that the earth received its magnetism from some such cause?

In short, may not a magnetic power exist throughout our system, perhaps through all systems, so that if men could make a voyage in the starry regions, a compass might be of use? And may not such universal magnetism, with its uniform direction, be serviceable in keeping the diurnal revolution of a planet more steady to the same axis?

Lastly, as the poles of magnets may be changed by the presence of stronger magnets, might not, in ancient times, the near passing of some large comet of greater magnetic power than this globe of ours have been a means of changing its poles, and thereby wracking and deranging its surface, placing in different regions the effect of centrifugal force, so as to raise the waters of the sea in some, while they were depressed in others?

Let me add another queſtion or two, not relating indeed to magnetiſm, but, however, to the theory of the earth.

Is not the finding of great quantities of ſhells and bones of animals, (natural to hot climates) in the cold ones of our preſent world, ſome proof that its poles have been changed? Is not the ſuppoſition that the poles have been changed, the eaſieſt way of accounting for the deluge, by getting rid of the old difficulty how to diſpoſe of its waters after it was over? Since if the poles were again to be changed, and placed in the preſent equator, the ſea would fall there about 15 miles in height, and riſe as much in the preſent polar regions; and the effect would be proportionable if the new poles were placed any where between the preſent and the equator.

Does not the apparent wrack of the ſurface of this globe, thrown up into long ridges of mountains, with ſtrata in various poſitions, make it probable, that its internal maſs is a fluid; but a fluid ſo denſe as to float the heavieſt of our ſubſtances? Do we know the limit of condenſation air is capable of? Suppoſing it to grow denſer *within* the ſurface, in the ſame proportion nearly as we find it does *without,* at what depth may it be equal in denſity with gold?

Can we eaſily conceive how the ſtrata of the earth could have been ſo deranged, if it had not been a mere ſhell ſupported by a heavier fluid? Would not ſuch a ſuppoſed internal fluid globe be immediately ſenſible of a change in the ſituation of the earth's axis, alter its form, and thereby burſt the ſhell, and throw up parts of it above the reſt? As if we would alter the poſition of the fluid contained in the ſhell of an egg, and place its longeſt diameter where the ſhorteſt now is, the ſhell muſt break; but would be much harder to break if the whole internal ſubſtance were as ſolid and hard as the ſhell.

Might not a wave by any means raiſed in this ſuppoſed internal ocean of extremely denſe fluid, raiſe in ſome degree

gree as it paſſes the preſent ſhell of incumbent earth, and break it in ſome places, as in earthquakes? And may not the progreſs of ſuch wave, and the diſorders it occaſions among the ſolids of the ſhell, account for the rumbling found being firſt heard at a diſtance, augmenting as it approaches, and gradually dying away as it proceeds? A circumſtance obſerved by the inhabitants of South-America in their laſt great earthquake, that noiſe coming from a place, ſome degrees north of Lima, and being traced by enquiry quite down to Buenos Ayres, proceeding regularly from North to South at the rate of—Leagues per minute, as I was informed by a very ingenious Peruvian whom I met with at Paris.

GEOLOGY

Thomas Cooper

Before I enter on the relative ages of metallic substances, it may not be amiss to give a brief sketch of the formation of the earth, according to the best acknowledged facts we possess, and the most probable opinions hitherto advanced. In doing this, I shall not scruple to blend my own views of the subject, with the remarks which I shall be induced to adopt from other writers, chiefly Werner, as exhibited by Jamieson; confining myself however, as well as I can, to fair deductions from known phenomena.

There are only two systems relating to the explanation of the general appearance of our planet, that are entitled to any consideration: the one the *Neptunian*, at the head of which is Werner the professor of mineralogy at Friburgh; the other the *Plutonian*, advanced by the late Dr. Hutton, so well known in the mathematical world, and at present chiefly supported by professor Playfair of Edinburgh.

Werner's system, in brief, is, that all the more extensive and universally-found strata, or formations, of our globe, have been formed, partly by crystallization of substances dissolved or intimately mixed with the watery fluid that contained them in a chaotic state—partly by subsidence of the particles mixed with the water—and in cases of volcanic strata, by volcanic eruptions. His general distinction of primitive, transition, secondary, alluvial, and volcanic soils or rocks, appears to me too probable to be rejected; nor is it possible for any person who has seen, (as may very commonly be seen,) granite and quartz; also plants and soft shells, surrounded by and enveloped in, limestone, flint, siliceous grit, and argillo-silite, to doubt, but the great majority of rocks and stones, are formed by crystallization and subsidence of particles dissolved or mixt in water.

According to the Plutonian hypothesis of Hutton and Playfair, our globe is subject to a gradual but perpetual change, inducing endless alterations of continent and of sea, in the same places. The present continents, for instance, are subject to destruction by the action of air, rain, mechanical attrition, chemical decomposition, the operation of gravity, &c. The materials thus broken down, and decomposed, are gradually carried to the bottom of the ocean, where they are subject to induration by the action of internal heat, and new strata are formed, which in time are raised by subterraneous fires, becoming in their turn terra firma. The sea is propelled over the old continents, which then become the bottom of the ocean, while the new continents are gradually clothed with vegetables and animals, and in process of time undergo the same gradual action, decay, and submersion, which their predecessors experienced.

The experiments of Sir James Hall, on the effects of heat modified by compression, have been made in pursuance and support of Hutton's theory; and he has certainly shewn, that crystallized forms of carbonat of lime may be produced under the joint operation of great heat and great compression, which are very similar to such as would be generally ascribed to a crystallization from watery fusion or admixture.

The view of the subject that at present occurs to me as most probable, is this:

The density of the earth, according to the calculations and observations of Sir Isaac Newton, Dr. Maskelyne, Dr. Hutton, Professor Playfair, and the honourable Mr. Cavendish, cannot be less than five times that of water.

The strata or formations of the earth, so far as they have been examined, consist of the following nine earths or their combinations: alumina, silica, calcia, magnesia, baryta, strontia, glucina, zirconia, yttria. The three last are found in quantities so comparatively small, as not to be worth notice on the present occasion. The same observation may be made, though in a less degree, on baryta and strontia, which are only found occasionally in secondary strata in nodules, or as the matrix of ores, or otherwise insulated. The earth and its formations may therefore be considered as consisting of alumina, silica, calcia, and magnesia, and their admixtures and combinations; interspersed rather than intersected occasionally by metallic substances.

Alumina has the specific gravity 2, calcia, 2.3, magnesia 2.3, silica 2.65. The metallic ores contained in these formations, are too few in number and in quantity to raise the specific gravity of the mass ,1. Add to this, that the quantity of sea & river water contained in these formations will greatly reduce the specific gravity of the mass; so that the average specific gravity of all the known strata of the earth cannot fairly be considered as amounting to more than 2; but if taken at 2.5, then will the known strata possess a specific gravity of one half the specific gravity of the whole globe. Hence it will follow, that this earth consists of a nucleus, of a metallic nature, whose specific gravity *exceeds* 5, covered by a crust consisting of a series of formations having together a specific gravity not quite reaching 2.5.

This outward crust, including the rivers and oceans that rest upon and within it, seems to be the only object of examination to the Geologist, or as the German philosophers affect to say, the Geognost. No observation that I know of, has hitherto extended beyond the granite formation that appears as its substratum. I know of no volcanic ejection that will warrant us in concluding that the matter thrown out, is any other than part of the formations that constitute this crust. The following questions admit but of conjectures.

What are the constituent parts of the nucleus?

Is there any series of cavities between the nucleus and the crust?

What is the thickness of the crust?

Of how many original and universal strata, or formations does this crust consist?

I tread upon new ground; but I use the aids which wise men have furnished; I have none of my own.

The nucleus is metallic: I conclude this, from its specific gravity; far too great for any known earth. Sir Isaac

Newton computed it at 7, but I abide by Dr. Maskelyne and Mr. Cavendish. If, according to them, the whole mass of the earth be 5, the nucleus cannot be less than 6, considering the deductions to be made for the various earths forming the crust, the waters that cover so large a part of it, and the cavities that are most probably contained in it.

Of what metallic substance? Probably of iron and nickel: because these are the only magnetic metals; and I know not how possibly to account for the phenomena of magnetism, but by means of a magnetic nucleus. Moreover, I cannot but suspect that some connection exists between the composition of meteorolites, all of which contain iron and nickel, and the subject of this investigation. The specific gravity of metallic nickel is only 8.38, of iron 7.6 or 7.7. The nucleus however will more probably consist of the ores of these metals than the metals themselves: and we know that very many of the iron ores are magnetic and polar. These are conjectures: but the present state of our knowledge does not afford better.

Are there any cavities intervening between the nucleus and the crust? Cavities which admit of the entrance of atmospheric air?

It should appear that there are such. For,

It is manifest that the whole series of formations from the uppermost alluvial soil down to the lowest granite, have in many instances been shaken en masse, from their foundations—upheaved. None of them are horizontal, as they were originally; a fact which Saussure first established: nor do any of them preserve an uniform dip or inclination. Marks of the revolutions they have undergone ab imo, from the very deep, are not to be gainsaid. The lowest and deepest granite, is most generally found also as the outgoing, or as constituting the summit of the highest mountains—breaking out to the day, in the well chosen phraseology of the miners. Nor is the older granite always in its original situation, undermost: it is found sometimes overlaying gneiss and other primitive strata. Evidently the result of eruptions: taking place, not in cavities that occur in or between the layers of the crust, but below them all; for they have all been raised up from the lowest granite, with its superincumbent formations by some deep seated and mighty force.

Will it not be allowed, that this force is probably volcanic? Whence otherwise is it to be derived?

Electrical earthquakes have had their day: they will occur no more. Nor shall we imitate (I suspect) the Neapolitan philosopher who proposed to sink wells to let out the steam of the great abyss. I venture to assert it, as a theory at least as likely as any other hitherto proposed, that volcanoes and earthquakes, are owing to the chemical action on each other, of iron, sulphur, moisture and atmospheric air. Where are these to be found?

All the sulphur of Europe, is supplied by the sublimation of that substance in the Solfaterras—in the immediate vicinity of European volcanoes.

All lavas contain (on the average) 24 per cent. of iron. I think this is the quantity which Kirwan states.

We have then as volcanic products, sulphur and iron; and if through the lower strata or formations, water should be supplied, we have even without atmospheric air, all the materials for earthquakes, volcanoes, eruptions and subversions. The water is decomposed: the iron and sulphur oxygenated: caloric evolved: hydrogen escaping through immense cavities inflamed, and all the phenomena at once accounted for.

Having now arrived, *per varios casus, per tot discrimina rerum,* from the nucleus to the crust of the earth, we will try what is to be done with that important part of our subject.

The following circumstances have been observed, and may be taken as facts. 1. There appears to be a series of strata, or as Werner calls them formations, that may be considered as surrounding the nucleus of the earth. The first formed, or lowest series, always preserve the same situation to each other except where occasional eruptions, or circumstances not of a general nature, make a variety in their situations. These strata are not only the deepest, but they are also the highest that are observed in the crust of the earth; forming the tops of the highest mountains. They are characterised by an appearance of crystallization, and by containing no remains of organic matter, vegetable or animal. The strata or formations that in general constitute this first, deepest, highest, and crystallized series, are,

Granite, consisting of feldspar crystallized in facets frequently lustrous; quartz; mica. Sometimes also schorl. Sometimes the schorl, sometimes the mica, sometimes both are wanting. But these are accidental deficiencies. This stone in all its varieties, is common about Baltimore, and at Germantown.

Gneiss.—This is a stone composed of feldspar, quartz, and mica, in much smaller particles than in granite; in the mass, it is also stratified or formed in layers, which granite is not. This is the common stone used for building, and for kirb stones in Philadelphia.

Mica Slate.—This is a stratum or formation consisting principally of quartz and mica, in which the mica predominates. It generally also contains crystallized garnets. Stones of this formation are common about Germantown, the Falls of Schuylkill, &c.

Clay Slate.—The common grey, bluish, yellowish, or smoke coloured slate, often used for covering houses.

Primitive Trap.—This is the pure black hornblende, the hornblende slate, and the mixed hornblende; it appears at 11 and 12 miles from Philadelphia on the Chesnut Hill road immediately after the steatite or soap stone; and intermingling with the micaceous shistose limestone, and then with the granular limestone.

Granular Limestone.—Crystallized: this may be observed on the road to White Marsh, about 13 miles from Philadelphia: the mica slate, becomes gradually micaceous limestone slate, and then granular limestone or marble, coloured with hornblende (amphibole, primitive trap) as in the black, and black and white marble of White Marsh used at Philadelphia. The clay slate here, does not intervene so far as I recollect.

Serpentine.—I have not traced this in the neighbourhood of Philadelphia: the soap stone first appears. This I think passes into serpentine, of which the neighbourhood of Easton furnishes fine specimens. Chlorite escaped me.

Porphyry and Sienite.—Porphyry, is compact feldspar, containing small crystals of feldspar: or quartzose stones containing such small crystals. Sienite, is a stone composed of feldspar, quartz and hornblende. I suspect this formation has not been traced upon, or over lying the serpentine in Pennsylvania. These stones abound, as rolled specimens out of place, on the shores of the North East branch of the Susquehanna from Danville upward to Wilkesbarre, and I believe at intervals as far as Tioga point. I know not the source of them. I suspect them to belong to the secondary trap formations, for they are intermixed with all the varieties of green stone. The feldspar is generally reddish.

Sometimes a second deposition of granite, or newer, more recently deposited granite, is found among the primitive strata: in this the crystals are similar to those of the oldest granite formations, but smaller. If I mistake not, Norristown will afford specimens of this.

These strata or formations are so generally found, and in the same situations as incumbent upon or subtending each other relatively, that they may be considered as universal.

Their crystallized appearance shews that their particles have been either dissolved or very finely suspended in water, so that the attraction of crystallization has been free to operate: that this water has been deep, so as that the lowermost parts of it have not been much agitated during the crystallization, which would otherwise have been more confused than it is: and indeed the oldest formations are the best crystallized. A part of the water covering the nucleus must have been taken up as water of crystallization, in the primitive formations. Again: when these were deposited, there were no vegetables formed: of course no animals: nay even the sea was unpeopled, for there is no trace of any organic remains in these strata. Even the Belemnites, the Asteriæ, the Echini, the Entrochi, the most simple forms of oceanic animal life, do not occur till the transition strata appear. Hence the propriety of denominating these formations, *primitive*.

By processes of nature, (beside the consumption of water by the new crystallized masses) to us unknown, the mass of waters appear to have diminished. The higher parts of the primitive strata or formations became the shores to the water superincumbent on their bases and middle regions; the simplest forms of oceanic animals came into existence, and mosses and lichens of high latitude, would generally occupy the surface of the primitive strata, gradually decomposed by the alternate action of air and water after many ages.

During this period, while the strata were in a state of *transition* from the chaotic to the habitable state, other deposits would gradually be made from the waters, now decreased in quantity; and take their place below the summits of the primitive range. Those summits being exposed to the action of the atmosphere, of rains, of frost probably, and to the action also of the waters with their contents still incumbent on the earliest strata, would furnish masses and particles washed away, which would mingle with the deposits of the transition series: this series therefore, will exhibit appearances of mechanical and chemical intermixture of earths and stones such as are found in the green stone, siliceous hornblende rock, argillaceous hornblende rock, grauwacky, and lastly, wacky which form the trap rocks of the transition series. Specimens of these trap formations can be traced from Perkiomen Bridge, through Reading to the mountain ranges of shistus that reach from Putt's forge to Sunbury, in Pennsylvania. The transition limestone is the earliest of this series, but I have not had occasion to remark the flint slate or the transition gypsum. During the period when these transition formations were deposited, there would be no land animals, for there would be no vegetables for them to feed upon. There would be no vegetables, unless some few lichens, mosses, or ericas, that would find foothold upon the slight decomposition that after the lapse of some ages would take place on the surface of the primitive rocks. The sea only would be peopled, and that but sparingly; for in that mass of muddy water, none but the lowest and most inferior grades of animal life, and such as do not inhabit deep water could exist. Hence we find the transition formations contain in their substances, some belemnites, asteriæ, entrochi, echini, &c. but no organized vegetable substance except very rarely in the latest rocks of this series, and no remains whatever of terrestrial animals.

Indeed, in the high latitudes of the outgoings or summits of the primitive strata, very few vegetables even at the present day can live. No vegetation fit for animal life, could take place, until the transition, and most of the next series of (secondary or fletz) formations had subsided. These would occupy gradually lower and lower situations, till a rich soil from every kind of intermixture of earth mechanically deposited, would afford a proper temperature of region, and an easily decomposed soil wherein vegetables could grow.

Next to the transition series, then, come the *secondary* or as the German mineralogists call them, the Floetz (Fletz) Rocks: so called, because they appear to be more floated or horizontal; though I confess the appellation does not appear to me peculiarly appropriate.

These strata, consist principally of sandstone, limestone, sometimes stinking from bituminous impregnation, sometimes shelly; secondary trap, graphite and bituminous coal, gypsum, rock salt.

The old red sandstone, limestone, secondary trap, and newer sandstone, are to be found in Adams and York counties. The graphite or *anthracite coal formation** in Pennsylvania, extends from the Delaware at the heads of Lacawana and Lehigh to the North mountain, whose southern base I think it subtends. It is found ten miles all round Wilkesbarre; it is found on the Berwick turnpike road; it is found a mile from the turnpike road 21 miles from Sunbury toward Reading; it is found on the Susquehanna, or within a few hundred yards of it, six miles below Sunbury; it is found at the iron works on the waters of the Schuylkill, on the road from Reading to Hamburgh†. This stratum does not extend westward beyond the west branch of Susquehanna. I do not know whether it is connected or not with the anthracite of

* Carbon with little sulphur and no bitumen.
† And b. Mahantongo 8½ miles from Susquehanna.

Rhode Island, not having been there. But this last seems to border on primitive formations, under circumstances, which for want of knowledge, I cannot explain.

The *bituminous coal formation* in Pennsylvania, exhibits its first trace as breaking out to the day, on the Juniata. The bed of the river at Chingleclamoose, up the west branch of Susquehanna, is bituminous coal. It extends from thence northward and westward throughout the whole of Pennsylvania. How far it extends on the Monongahela, Alleghany, and Ohio, I cannot say. The shell limestone extends up Sugar creek on the west side of the north east branch of Susqa. toward the heads of that creek, where it abounds so much in shells as to be fit for lime. All the stones about the Sheshequin abound in shells. This stratum I have traced downward (south westward) through Buffalo to Jacks' mountain, which is a mass of shell lime and calcareous Breccia. How far it extends in both directions I know not.

I have observed that the Alpine heights of the primitive mountains could at no time furnish much vegetable food. The same remark, but in a less degree, will apply to the transition range. The low and kindly climates occupied by the secondary series—the soft and decomposable nature of these depositions would furnish the true theatre of vegetable life: and until these regions were filled with vegetables, the race of animals could not have been produced; for on what could they subsist? Gramenivorous animals therefore must have succeeded the various forms of vegetable existence, and carnivorous the gramenivorous.

The vegetable matter imbedded in the substance of the secondary strata, will consist of the remains of vegetables that grow on the transition strata; and the animal remains will consist chiefly of such animals as were produced in early stages of animal existence, particularly the smaller aquatic animals; and of these, chiefly shell fish, as shells are not so soon decomposed as mere animal substance.

Coal and bituminous substances, will also occur in the latter floetz or secondary formations only; for with me there is no doubt of these being the produce of submerged vegetables, subjected to the effects of heat modified by compression. The many specimens I have seen at Whitehaven and elsewhere in England, and some American specimens that I possess myself, wherein there is an evident gradation and passage from wood to coal in the same piece, compels me to adopt this opinion. The process of nature in converting wood into coal, I do not pretend to have satisfied myself about.

The latest deposits of what is considered as the secondary series of formations, will comprehend basalt, wacke, greystone, amygdaloid, newer limestone, chalk and calk sinter, obsidian, pumice, pitch stone,* (I have found this among primitive rocks at the mills near Baltimore) coal, gypsum and rock salt, which two last usually keep company in this series; argillaceous iron ore, petrifactions, &c.

The remains of land animals are nearly confined to the newest floetz and alluvial deposits. The remains of many land animals, have been found, particularly by Cuvier, of which the race does not now exist. So La Marck has found remains of aquatic animals similarly circumstanced. The same remark will apply also to vegetable remains.

It is evident that the newest of each series of formations will touch upon, and in some degree intermingle with the oldest of the succeeding series, and partake in some degree of the mutual characters.

As the waters gradually decreased, and retired into their oceanic basons, the finer kinds of earthy matter would subside the last: and those saline substances that

* I doubt about Pitch stone belonging to this series.

could no longer find fluid to hold them in solution, would crystallize.

These last are *alluvial* deposits; which also are sufficiently general to merit the title of formations, although varying in their composition perpetually, as clay, loam, marle, bog ore, sand, gravel, peat, &c. and their combinations. These occupy the lowest levels and the bases of the other rocks.

Beside these, occasionally we meet with *volcanic* rocks or stones; lavas. These contain debris or broken fragments of many deep formations, as the granite, mica slate, greenstone, hornblende, and sandstone found in ejected granular or primitive limestone. Hence some volcanic caverns are probably situated between the nucleus and the primitive strata: a situation which we are led to presume, from the inclination or dip of the oldest formations: their position would naturally have been level and horizontal, had not some mighty force raised them from their base. What effect the influence of the moon has had, in determining the circumstances of these earthy depositions out of the immense body of water that in their chaotic state contained them, no one now can fully explain: that it must have produced oceanic tides then, as well as now, perpetually varying, with the varying density of the mass of turbid fluid acted upon, we can hardly doubt. It is evident also that many ages must have passed before the surface of our globe, put off its chaotic state, and became fit for the habitation of man. The *general* system, of which I have presented a scanty outline, seems pointed out in most of its parts, by facts and appearances not to be denied, or by other theories so well explained: but as to all the *particulars*, doubts and uncertainties hang over them, which more accurate and future observations may in some degree serve to explain.

On the calculation of the oblateness of the earth, by means of the observed lengths of a pendulum in different latitudes, according to the method given by La Place in the second volume of his "Mécanique Céleste," with remarks on other parts of the same work, relating to the figure of the earth.

BY NATHANIEL BOWDITCH, LL. D.

Upon looking over Dr. Rees' Cyclopedia, under the article *Earth*, I found he had inserted the elegant method of computing the oblateness of the Earth, or the ratio between the polar and equatorial diameters, by means of the observed lengths of a pendulum vibrating in a second of time in different latitudes, as it was published by La Place in the second volume of his immortal work the "*Mécanique Céleste;*" but he has allowed the application of the formulas to numbers to remain nearly as in the original work,[*] and has moreover committed some mistakes of his own, so that the article, as it now stands, is quite imperfect; and as this Cyclopedia has an extensive circulation in our country, it seems to be proper to notice these errors, in order that a currency may not be given to inaccurate ideas on the subject. It may also be mentioned as an additional reason for noticing them, that by correcting the

---

[*] In this work there is a mistake which was alluded to in a note to my paper on the total eclipse of 1806, published in the Memoirs of the Academy, in the year 1809, vol. 3, p. 21.

calculation we obtain for the oblateness of the earth, a result much more conformable to those deduced from other methods; and on this account I have thought it would not be unacceptable to the Academy to have these mistakes corrected, and the sources of them pointed out. This is done in the *first* section of the present paper. In the *second* section I have simplified one of La Place's formulas relative to the figure of the earth. In the *third* section I have corrected the expressions of the length of a degree, and also the azimuths given by him in § 38, Book III, of his "*Mécanique Céleste*," (in the hypothesis that the earth is not a spheroid of revolution) for the mistakes arising from the neglect of one of the terms in the expression of the radius of the earth, which produces a considerable effect in the value of one of the formulas.

SECTION FIRST.

There are *four* methods generally used for the purpose of computing the oblateness of the earth, supposing it to be an ellipsoid of revolution. *First.* By comparing the observed lengths of two consecutive degrees of the meridian. *Second.* By comparing the lengths of two degrees of the meridian measured in very different latitudes. *Third.* By means of the observed variations in the lengths of pendulums vibrating in a second of time in different latitudes. *Fourth.* By means of two equations in the moon's motion (the one in longitude the other in latitude) depending on the oblateness of the earth.

The *first* method is liable to much uncertainty. For the greatest difference between the lengths of two consecutive degrees, being only 9 or 10 fathoms, the least error in the lengths of the lines, in the observed angles, or in the altitudes of the heavenly bodies by which the length of the celestial one is determined would produce a great error in the computed oblateness. This was the case with the first observations made in France, by Picard, Cassini, &c. for the purpose of determining the form of the earth. The errors of the observations affected the result so much that it was supposed by many Mathematicians that its figure was prolate, or lengthened in the direction of the polar axis. Even the late very accurate measures made in France by Messrs. Delambre and Mechain, and in England by General Roy, make the oblateness of that portion of the earth nearly double what its general value is found to be by more distant observations. The greater part of this difference no doubt arises from a real irregularity in the figures of the meridians of the earth; but the method of computation itself labours under a similar defect to that which exists in finding the distance of any terrestrial object by means of its azimuths observed from the extremities of a base line, whose length is very small in comparison with the observed distance of the object. The *second method* of determining the earth's figure by means of distant observations is much more accurate than the preceding, but the various errors of the observations, and the irregularities of the surface of the earth, have a very perceptible effect on the oblateness computed by this method. These irregularities ought not to be so sensible in the results of the *third method* by means of the observed lengths of pendulums, as La Place has proved in Book II, § 33, of his "*Mécanique Céleste*," and with respect to the *fourth method*, it is evident without any calculation that it must be almost wholly independent of this same error. For, on account of the great distance of the moon from the earth, the effect of all the little irregularities of the earth's form, will be hardly perceptible in the attraction on the moon, and the result of the general figure only will prevail. Therefore by using the last method and taking a great number of observations, it would seem that we ought to obtain a more correct value than by any other way. The next method in point of accuracy would be by the observed lengths of pendulums, and that depending on the actual measures of the degrees of the meridians, which at first sight appears to be the most natural and accurate, would be in fact the least accurate of any of the methods here mentioned.

To obtain the oblateness by means of the lunar equations, the indefatigable Astronomer Burg undertook to compute the coefficients of these equations by means of such of the observations of Maskelyne, as were proper for that purpose, and from these values La Place has computed* the oblateness of the earth. The equation in longitude made it $\frac{1}{305.05}$ and the equation in latitude $\frac{1}{304.6}$, which differ from each other but a very small fraction, and this wonderful agreement of two independent calculations is a great proof in favour of the accuracy of the method, and of the correctness of the result; and if we neglect the decimal parts of these numbers, we may put $\frac{1}{305}$ for the oblateness as determined by this method, and in all probability this is very near to its correct value; we should therefore be led to infer from what has been said, that the oblateness determined by the observations on the pendulums would agree very nearly with this quantity; but according to La Place's calculation this is not the case. For by combining all the observations of the pendulums which he considers as sufficiently correct, he has found the oblateness to be $\frac{1}{335.78}$,† which differs

---

* In Book VII, § 24, 25.

† This is given in Book III, § 42 of his "*Mécanique Céleste.*" Making use of the observations in Peru, Porto-Bello, Pondicherry, Jamaica, Petit-Goave, Cape of Good Hope, Toulouse, Vienna, Paris, Gotha, London, Petersburgh, Arensgberg, Ponoi, and Lapland.

considerably from the preceding. On the other hand by combining the measured degrees of the meridian, in a similar manner, he finds the oblateness to be $\frac{1}{312}$ *; which agrees much better with $\frac{1}{305}$ than the quantity $\frac{1}{335}$ obtained by the pendulums, directly contrary to what it ought to be by the theory, which teaches us that the method by pendulums ought to be more accurate than by the measures of the degrees of the meridian. This difference will be in part rectified if we use the corrected measure of the degree of Lapland as found by Svanberg, but even with this correction the method of pendulums will differ the most from $\frac{1}{305}$ which we have supposed to be the correctest value. This difficulty will, however, be wholly obviated if we correct the calculation for the two numerical mistakes mentioned above, by which means the oblateness will be increased from $\frac{1}{335.78}$ to $\frac{1}{314.75}$ and then the results of the methods by the lunar equations, the pendulums, and the degrees of the meridian will be respectively $\frac{1}{305}, \frac{1}{314.7}$ and $\frac{1}{324}$, which are in the natural order indicated by the theory.

[*Editor's Note:* The remainder of this article (pp. 35–49) contains details of Bowditch's recalculations of the La Place equations and has been omitted here.]

---

* This is given in Book III, § 41, of his "*Mécanique Céleste*," using the degrees measured at the Equator, Cape of Good Hope, Pennsylvania, Italy by Boscovich and Le Maire, in France by Delambre and Mechain, in Austria by Liesganig, and in Lapland by Maupertuis. If we use the corrected measure of the degree of Lapland as found by Svanberg and given in Rees' Cyclopedia namely 57196 toises for the latitude 66° 20′ 10″. The oblateness will be decreased from $\frac{1}{312}$ to nearly $\frac{1}{324}$. I took its length of the whole arch measured equal to 1.5°. as given by Dr. Rees, as I had not Svanberg's work to refer to.

Part VI

FORMATION AND CLASSIFICATION OF ROCKS

Editor's Comments on Papers 17 Through 22

17 **SMITH**
Account of Chrystallized Basaltes Found in Pennsylvania

18A **HALL**
An Account of a Supposed Artificial Wall, Discovered Under the Surface of the Earth in North Carolina

18B **WOODHOUSE**
Reply

19 **LEWIS**
Remarks on a Subterranean Wall in North Carolina

20 **MACLURE**
Excerpt from *Essay on the Formation of Rocks, or an Inquiry into the Probable Origin of Their Present Form and Structure*

21 **MACLURE**
Hints on Some of the Outlines of Geological Arrangements, with Particular Reference to the System of Werner, and with Introductory Remarks by Benjamin Silliman

22 **DAY**
A View of the Theories Which Have Been Proposed, to Explain the Origin of Meteoric Stones

The origin of rocks, their classification, and the science of geology are intimately connected. Geology is a historical science, and an understanding of the origin of rocks is an essential element of that history. Rock classifications are primarily generic and consequently depend on theories of formation. It is therefore hardly surprising that progress in understanding the formation of rocks, rock classification and geology itself should be closely tied.

Abraham Gottlob Werner (1749–1817), gifted teacher at the

Freiberg Mining Academy and one of the founders of modern geology, developed the system of rock classification that was universally accepted in America before 1800. The Wernerian system, as outlined by Thomas Cooper (Paper 15, pp. 418-425), was based on the theory that the vast majority of crustal rocks formed as precipitates from a vast global ocean. The sequence of crystallization was hypothesized to be the same in all parts of this sea, thus simplifying the task of the field geologist. The proposed sequence began with *primitive* or *primary* rocks, including granites, gneisses, and other nonfossiliferous crystalline formations. Cooper explained that "their crystallized appearance shews that their particles have been either dissolved or very finely suspended in water." Following deposition of the primitive rocks, which underlay all other crustal deposits, the ocean receded in places, exposing the primitive strata to air, and *transition* formations began to appear. These precipitates were the first to contain life and were intermingled with erosion products from the exposed primitive lands. *Secondary* or *Flötz* rocks were younger sediments that in places contained the first signs of terrestrial life. Cooper cites the coal deposits in Pennsylvania as an important secondary formation of the North American continent. Receding waters of the once-continuous ocean left the final *alluvial formations,* which were superficial to all others.

Werner's ideas were championed in the English language by Robert Jameson of Edinburgh, whose geology text reached a wide and attentive audience. The clarity and simplicity of Werner's system attracted many adherents in both Britain and America, and the aqueous origin of most rocks appeared obvious to the trained eyes of these earth scientists. A typical account is that of Thomas P. Smith (Paper 17), who observed basalt in crystal-like jointed blocks. These regular forms he took to be "strong corroborating proof of their Neptunian origin." Because deposition in the Wernerian ocean was uniform, Smith suggested that a careful search of adjacent hills should reveal similar deposits of basalt.

The difficulty in assigning unambiguous origins to rocks is well illustrated by a series of articles that appeared in the *Medical Repository.* A small trap dike in North Carolina weathered out from adjacent country rock so as to resemble a wall of human construction. The Reverend James Hall (of no apparent relationship to either the New York State paleontologist or the British geologist who had the same name) described the phenomenon to University of Pennsylvania chemistry professor James Woodhouse in a letter of 1798 (Paper 18A). Hall interpreted the find as a man-made production of great antiquity. As local investigators dug deeper along the sides of the "wall," they reached solid rock on either side, thus indicating that the wall was built before the country rock had hardened.

James Woodhouse (1770-1809), one of America's first chemists and an active investigator of minerals and their compositions, recognized in Hall's descriptions and accompanying specimens the characteristics of a regularly jointed basaltic dike (Paper 18B). Chemical tests on the rock added evidence to support this view. A dike was difficult to explain on the basis of aqueous deposition, and Woodhouse suggested that the rocks were volcanic productions, adding, "That there have been volcanoes in North Carolina appears from some specimens of lava sent from that part."

Woodhouse's analysis was not the last word on the North Carolina wall, for a Connecticut clergyman, Zechariah Lewis, visited the formation in 1800 and reinterpreted its history (Paper 19). After a careful examination, Lewis concluded that the wall was certainly a product of human workmanship. Sources for the stone and cement were identified, and possible motivations for its construction proposed. The true origin of the dike was still in question in 1822 when John Beckwith published his account of the phenomenon in the *American Journal of Science.*

As the body of field data increased, it became more and more evident that Werner's sequence of aqueous deposition was not universally applicable. James Hutton of Scotland challenged the Wernerian system with the theory that many rocks are created through the agency of heat and not water. Americans soon realized that a combination of these theories provided the best system of classification. First Parker Cleaveland (1816) and then William Maclure (Paper 20) adopted a mixed system that considered both aqueous and caloric agents in the creation of rocks.

William Maclure (1763–1840) was the dominant figure in American geology in the decade before 1820, primarily because of his epic *Observations on the Geology of the United States,* the first large-scale field study by an American. Merrill (1924) called American geology before 1820 the "Maclurean Era," and virtually all American writers on geology from 1818 to 1830 acknowledged their debt to Maclure. He was born in Scotland, died in Mexico, and spent many years living and traveling in Britain, Spain, and other parts of the world. He made his fortune in commerce at an early age and devoted many of his most productive years to American field geology. As a philanthropist, Maclure attempted to aid the educational system of Spain by establishing an agricultural school, and he endowed and participated in the communal experiment at New Harmony, Indiana. William Maclure was a generous, gifted, and pivotal personality in the history of American geology.

Maclure's "Essay on the formation of rocks" (Paper 20) is a carefully reasoned statement regarding the origin of rocks and their classification. All authorities, he began, acknowledged that the earth formed from a fluid state, but they disagreed regarding the relative importance of water and heat in creating and maintaining that fluid. (The originally fluid state of the spinning globe he proposed as the origin of the earth's oblateness.) Rather than confining himself to just water or fire, Maclure proposed two classes of rock, the first for Neptunian and the second for Plutonic rocks. A third class, for rocks of unknown origin, was also proposed, but it was hoped that this class would eventually be eliminated.* Maclure is sometimes cited as an advocate of the Wernerian system because the nomenclature in the *Observations* was that of Werner (however, see White 1970). It is clear, however, that Maclure's use of the terms "primitive," "transition," "secondary," and "alluvial" was due more to the general familiarity of his readers with those names than to an endorsement of the details of Werner's theory. In his "Essay on the formation of rocks," Maclure abandoned much of the Wernerian nomenclature and presented a balanced and practical solution to the problem of classifying rocks.

The clearest indication prior to 1820 of Maclure's growing dissatisfaction with some aspects of Wernerian classification was contained in a brief note in the first volume of the *American Journal of Science* (Paper 21). Maclure's "Hints on some of the outlines of geological arrangements" is prefaced with an extended editorial note by Benjamin Silliman (1779–1864), founder of the *American Journal* and distinguished professor of natural sciences at Yale University. At the time of his appointment, the young Silliman had no training in science. Yale President Timothy Dwight could find no qualified teachers of chemistry and mineralogy and decided to send Silliman, an excellent law student, to Princeton, Philadelphia, and Europe, to gain experience in these sciences. Needless to say, Silliman excelled in his studies and became a leader in the teaching and publication of American science for many years.

In his preface to Maclure's letter, Silliman gave an admirable definition of the empirical science of geology:

> Geology . . . means a not merely theoretical and usually a visionary and baseless speculation, concerning the origin of the globe; but, on the contrary, the result of actual examination into the nature, structure, and arrangement of the materials of which it is composed.

Maclure, Silliman continued, was uniquely qualified as a geologist, for:

> Few men have seen so much of the structure of our globe . . . . His opinions on some of the more obscure and doubtful parts of the Wernerian geology are worthy of peculiar consideration, for they are founded on a course of observations vastly more extensive than Werner ever had in his power to make.

---

*Maclure's description of his own classification system is in error at the bottom of p. 266. There are three classes in the system rather than the two mentioned in the text at this point.

Editor's Comments on Papers 17 Through 22

Silliman concluded that, although Werner would always be remembered with esteem as a key figure in the development of geology, his system must be subject to revision based on new observations.

It was with this preface that Maclure's letter questioning aspects of Wernerian theory appeared. Maclure's "Hints" related in part to specific rock types in the vicinity of New Haven, Connecticut, but a more general theme was sounded. Maclure objected to several aspects of Werner's nomenclature, noting that "misapplication of names naturally arises (in some cases) from the system of neptunian origin." A more significant objection to Werner's theory was that secondary formations were observed to occur in *different* sequences in different localities, rather than in the universal sequence suggested by the German geologist.

> In geology the best mode for the greatest part of the secondary would be to give the relative positions of the strata of each valley or basin; and I am rather of the opinion that they would all differ from one another.

Thus Maclure demonstrated his reliance on empirics rather than speculation in describing the earth's crust.

The origins of meteorites or "meteoric stones" was a baffling mystery to researchers of the eighteenth and early nineteenth centuries. The widely viewed fall of a group of meteorites near Weston, Connecticut, in 1806 was a matter of wonder and amazement, rendered even more significant by the systematic description of the phenomenon by Yale professors Benjamin Silliman and James Kingsley (1808). The two Yale scientists analyzed the fall in a quantitative manner that convinced many readers of the existence of meteorites. There were, however, some notable skeptics including Thomas Jefferson who is reputed to have quipped, "It is easier to believe that two Yankee professors could lie than to admit that stones could fall from heaven" (Fulton and Thomson 1947).

An 1810 reprint of the Silliman and Kingsley account was followed by a more general exposition on the origin of meteorites by the Yale mathematics professor Jeremiah Day (Paper 22). Day reviewed four possible origins of meteors. He all but dismissed the possibilities that meteorites are atmospheric condensates or volcanic productions from the earth or moon. Day favored a fourth hypothesis, "though not entirely without difficulties." "Terrestrial comets" in very eccentric *earth* orbits were presumed to be the objects that periodically fell to earth. The principal difficulty attending this explanation was why the stones were hot after landing. The friction of air seemed insufficient to generate the high temperatures of recorded falls. The origins of extraterrestrial rocks were thus as difficult to deduce as those of the earth.

REFERENCES

Cleaveland, P. 1816. *An Elementary Treatise on Mineralogy and Geology.* Boston: Cummings and Hilliard.

Fulton, J. F., and E. H. Thomson. 1947. *Benjamin Silliman—Pathfinder in American Science.* New York: Schumann.

Merrill, G. P. 1924. *The First One Hundred Years of American Geology.* New Haven, Conn.: Yale Univ. Press.

Silliman, B., and J. Kingsley. 1808. Account of a Remarkable Fall of Meteoric Stones, in Connecticut. *Phila. Med. Phys. Mus.* **3**:39-57.

White, G. W. 1970. William Maclure Was a Uniformitarian and Not a Real Wernerian. *Jour. Geol. Education* **18**:127-128.

BIBLIOGRAPHY

Beckwith, J. 1822. A Memoir on the Natural Walls, or Solid Dikes, in the State of North-Carolina; About Which There Have Been Debates, Whether They Were Basaltic, or of Some Other Formation. *Am. Jour. Sci.* **5**:1-7.

Brickell, J. 1809. On the Subject of the Falling of Stones from the Atmosphere. *Baltimore Med. Phys. Rec.* **2**:55-58.

Holly, H. 1808. An Investigation of the Facts Relative to the Descent of Stones from the Atmosphere to the Earth, on the 14th of December, 1807, in the Towns of Fairfield, Weston, and Huntington, Connecticut, and to the Meteor Whence These Earthy Bodies Proceeded. *Med. Repos.* **11**:418-421.

Latrobe, B. H. 1809. An Account of the Freestone Quarries on the Potomac and Rappahannock Rivers. *Am. Philos. Soc. Trans.* **6**:283-293.

Lewis, Z. 1802. Letter on Subterranean Wall on the Yadkin in North Carolina. *Med. Repos.* **5**:397-407.

Stevens, J. W. 1799. On the Creation and Convulsions of the Earth; The Inequality of Its Surface, and the Diminution of the Sea. *Weekly Mag.* **4**:225-232.

Woodhouse, James. 1804. Additional Observations on the Subterranean Minerals Near the Yadkin, and on Their Basaltic Nature. *Med. Repos.* **7**:26-27.

Woodhouse, James. 1808. Account of the Meteor which Was Seen at Weston, in the State of Connecticut, on the 14th of December, 1807; With an Analysis of the Stone. *Phila. Med. Mus.* **5**:131-133.

Account of CHRYSTALLIZED BASALTES *found in Pennsylvania—By* THOMAS P. SMITH.

Read, Jan. 18, 1799.

THE first place at which I found these basaltes was on the Conewaga hills, east of the Susquehanna and about half a mile to the north of Elizabeth-town. They are here to be found in considerable quantities, both chrystallized and amorphous—The chrystals are generally tetrahedral and of a very fine grain. There are great masses of it lying about amorphous, but it generally has a very strong apparent tendency to chrystallize.—As I travelled in the stage I had not an opportunity of examining this place as minutely as I could have wished. It is I think well worthy the attention of a minerallurgist whose time will permit and talents enable him to explore it accurately.

On my return from Northumberland by a different rout, I again found them at Campbell's town; they are here evidently a lateral branch of the Conewaga hills, and are scattered on the surface in the greatest profusion.

Soon after this I met with them on recrossing the Conewaga hills at Grubb's mines: as I now travelled in a private carriage, I had a much better opportunity of examining this part than near Elizabeth-town. At the foot of these hills Dr. Barton found a great quantity of regularly chrystallized *granite*, the predominate figure tetrahedral, higher up the chrystallized basaltes appeared; and what in my opinion is a strong corroborating proof of their Neptunian origin, they were interspersed with large masses of *brechia* composed of *silicious* pebbles evidently rounded by friction imbedded in the red free-stone of our mountains.

From

From thefe facts I am induced to believe, that if this chain of hills was accurately explored, it would be found to abound in its whole extent in chryftallized *bafaltes*, and in this opinion I am ftill further confirmed from having obferved the ftrong tendency the *whin*, as it is commonly called, has to affume a regular figure on a fpur of thefe hills I croffed in going from Lancafter to Columbia.

18 A

Reprinted from *Med. Repos.* 2:272-274 (1799)

AN ACCOUNT OF A SUPPOSED ARTIFICIAL WALL, DISCOVERED UNDER THE SURFACE OF THE EARTH IN NORTH CAROLINA

Reverend James Hall

SIR, *Philadelphia, May* 25, 1798.

NEAR the confluence of South Yadkin and Third Creek, about fourteen miles from Salisbury, in North-Carolina, a phenomenon of great antiquity has been discovered, which has engaged the attention of the curious in that part of the State, and which I have lately endeavoured to explore.

During the heavy rains which fell in the summer of 1794, a cavern of about eight feet deep was formed in the side of a hill, near a small stream of water, by the successive torrents of rainwater which issued from an adjacent field.

The hill is between two and three poles in surface where the cavern is formed, about the middle of which stands a subterranean wall, composed of small stones, laid in a white cement, resembling lime of a very fine texture. The largest stones, among many hundreds which I have examined, do not, in my opinion, exceed twelve pounds in weight, and from that are to be found of all sizes down to the weight of one ounce.

The species of stone is what the Irish call the *black whin;* nor is any other kind of stone to be found in the wall.

The stones incline to an oblong, though they are very irregular in their form.

They are universally laid across the wall; and the angles of each stone are so fitted by those contiguous to it, that it is difficult to enter the edge of a mattock between them, although the cement has lost its tenacious quality, and is as moist as the surrounding earth.

The cement I have examined in not less than forty different places in the wall, and could not find among it any appearance of sand, or common earth, except where there has been an opportunity of the earth mingling with it from the top of the wall.

Both sides of the wall are plaistered with the cement, so that not a stone has appeared when the wall was completed, supposing it to be a work of art; and that this is the case is, in my opinion, evident from this circumstance; that where the cement was washed off from the surface of the wall, hundreds of small stones appeared within a small space, as if slipped in between the ends of the stones, where they could not be brought into contact, to fill up the chasms with the cement about them; and from the apparent nature and situation of the materials, it appears probable to me, that when the wall was dry, and above ground, it was nearly as firm as a solid rock of the same dimensions.

The wall is about two feet thick, built in a straight line, and perpendicular. It has been traced about ninety feet down the stream below the cavern, and, perhaps, double that distance in the opposite direction. That part I did not measure.

The top of the wall, at an average, is between two and three feet below the surface of the earth, both above and below the cavern, although the situation of the ground on the two sides is very different.

Above the cavern the hill rises abruptly, where the wall rises with it, and, in a few poles, the ground becomes almost level, the stream bearing considerably from the wall.

Below the cavern the wall runs parallel to, and along the declivity of the hill; so that, as far as the wall has been explored, the end which is up the stream is, by a horizontal level, fifteen, perhaps twenty, feet higher than the other.

Where the wall bends over the hill the stones lie in a much more detached situation than in any other place which I examined, and appear as if the lower end had sunk when the cement was in a state of moisture, so as to admit the stones to be drawn asunder, rather than make any particular chasm.

Two circumstances have much excited my curiosity respecting the wall: one is, that it has been explored five feet lower than the surface of the adjacent stream, which runs not more than forty feet distant from the wall, without any appearance of its termination downwards, or of any end, corner or opening discovered in the distance of near three hundred feet in length where it has been traced. The other is, that a coarse gravelly rock embraced the wall on both sides, increasing in hardness as far as the wall has been examined in depth; from which I think these two facts are evident, that, at the time the wall was built, the adjacent stream had no existence in that place; and that, since that time, the rock has been generated.

The wall has had so little tendency to form a concretion with the rock, that, as far downwards as the rock would yield to the mattock where it was dug away, the plaistering stood smooth and entire; and, where the rock is too hard to be dug, and the wall was removed, the cheeks of the rock which embrace the wall are as level and smooth as the plaistering against which it rested. This, I think, incontestibly proves, that the rock has been formed after the wall was constructed.

The depth at which the wall has been examined is supposed to be about fourteen feet. This I could not exactly ascertain, as the wall had been demolished for near sixty feet in length.

It is my intention, if health permit, next August, to endeavour to carry forward a further inquiry, the result of which you may expect by the first convenient opportunity.

I am, Sir,
With much esteem,
Your obedient servant,
JAMES HALL, jun.

Dr. Woodhouse.

REPLY

James Woodhouse, M.D.
Professor of Chemistry
in the University of Pennsylvania

Sir,

I HAVE read your account of a supposed artificial wall, discovered under the surface of the earth, in North-Carolina, with great attention.

I am well satisfied, from several specimens of the stones which I have seen composing this wall, that it consists of a mineral substance called basaltes, and that it is a production of nature, and not of art.

My reasons for this opinion are as follow:

The stones answer the description of basaltes given by various writers. They are found of an irregular form, in prisms consisting of several sides, and are of different sizes; some being so small as to weigh no more than one ounce, while others exceed the weight of twelve pounds. The angles fit each other exactly like the basaltes, and appear as if joined by the hand of a skilful workman.

There is a brown ochreous matter found upon the surfaces of these stones, exactly like that on some of the basaltes of other countries. This ochre arises from a chemical decomposition of the stone, called by some spontaneous calcination, and by others efflorescence.

The decomposition is owing to the iron contained in the stones, and its calcination by air and water.*

Fourcroy has improperly attributed the brown crust with which the stones are covered, to water depositing different kinds of earth between the sides of the basaltic columns; and in Nicholson's Chemical Dictionary† it is called cement with equal impropriety. Columns of the Giant's Causeway, says the compiler of the Dictionary, fit accurately together, being, in some instances, united by a strong cement.

That the brown crust which adheres to the stones, and the fine white friable matter with which you suppose the wall has been

---

* Il eſt facile de voir que cette decompoſition, cette friabilité dependent du fer qui eſt contenu dans ces pierres, et de ſon oxidation par l'air et par l'eau. French Encyclopædia, art. bafaltes.

† Art. bafaltes.

plaistered, are owing to chemical decomposition, appears evident from the following circumstances: If the brown ochre is carefully scraped off from the stone, the surface will be found to be not of so firm a texture as the internal part; and the white powder, brown crust, and internal part of the stone, are composed of the same principle, in nearly the same proportions.

In some countries the basaltes are so much calcined as to fall to pieces on being removed.

The regularity of the wall, and the number of small stones which appear as if slipped in between the ends of the stones, are no proofs of its being a production of art. The basaltes, in Italy, appear like piles of wood of equal thickness throughout, and extend to a considerable distance. The small stones may have been carried down from the surface of the earth by rain, and deposited in the places where they are now found.

I do not suppose that the rock which embraces the wall, and which, from the specimen you have shewn me, is granite, was formed after the wall, granite being among the first formed substances in nature. The rock has probably been burst asunder by the wall, which is, perhaps, of volcanic origin.

In Cronstedt's Mineralogy there is an account, by Mr. Latrobe, of a rock of granite in Upper Lusatia, which has been rent asunder by a vein of concentric basaltes. In Italy basaltes are often found resting upon a bed of granite.

That there have been volcanos in North-Carolina appears from some specimens of lava sent from that part to this city.

The following experiments were made in order to ascertain the component parts of the American basaltes:

EXPERIMENT I.

One hundred grains of the solid stone were reduced into an impalpable powder, and boiled half an hour in half an ounce of nitric acid, diluted with one ounce of water. The whole was placed upon a filter, and distilled water was added until it passed through the filter—insipid to the taste. The powder remaining upon the filter was siliceous earth, and, when dry, weighed exactly fifty-eight grains.

EXPERIMENT II.

A solution of pot-ash was added to the fluid which passed through the filter until no precipitation took place. The precipitated matter was carefully washed in a large quantity of distilled water, and, when dried, weighed forty grains.

EXPERIMENT III.

This dried precipitate was boiled half an hour in distilled vinegar, in order to dissolve the lime and magnesia which it might contain. The vinegar was filtered and evaporated to dryness. Diluted sulphuric acid was added to the dry matter, in order to form selenite, or the sulphate of lime, and Epsom salt, or the sulphate of magnesia. Distilled water was added to separate the sulphate of magnesia from the insoluble sulphate of lime.

The magnesia was precipitated by a solution of pot-ash, and, when dried, weighed three grains.

EXPERIMENT IV.

That part of the dried precipitate, mentioned in the second experiment, which was not acted upon by the vinegar, weighed twenty-nine grains. It was dissolved in diluted nitric acid, and a solution of the prussiate of pot-ash was added until no precipitation took place. The prussiate of iron was separated by a filter, boiled in a solution of pot-ash, washed well with distilled water, and dried, when it weighed ten grains.

EXPERIMENT V.

A solution of pot-ash was added to the filtered liquor of the last experiment, until no precipitation took place. The precipitate, which was alumine, was well washed in distilled water, and, when dry, weighed sixteen grains.

The proportions of the ingredients composing the American basaltes, from these experiments, are fifty-eight parts of siliceous earth, sixteen of argillaceous, three of magnesia, and ten of iron, which, added together, make eighty-seven. Counting two grains lost in the first experiment, and five in the other, we will have ninety-four grains, which, with six allowed for the lime, will make one hundred grains.

One hundred grains of the white friable matter called cement, and the same quantity of the ochreous crust, were subjected to the same kind of experiments, and gave the following result:

	Silex.	Alumine.	Lime.	Mag.	Iron.	Loss.
White friable powder,	55	16	5	3	12	9
Brown ochreous crust,	54	15	6	3	11	11
Powdered stone,	58	16	6	3	10	7

Upon comparing this analysis with those of Bergman, Mongez, and Faujas de Saint Fond, no great difference will be found in the proportion of the ingredients composing the American basaltes and those of other countries.

Analysis by Bergman, Mongez, Faujas de Saint Fond:

Silex,	52	56	46
Argillaceous earth,	15	15	30
Lime,	8	4	10
Iron,	25	25	8
			6 Magnesia.
	100	100	100

This wall is certainly a great curiosity, and will afford ample room for the speculation of philosophers.

I should be happy to receive any further information upon the subject.

I am, Sir, with the greatest respect,

Your most obedient and humble servant,

JAMES WOODHOUSE.

Rev. JAMES HALL.
  June 1, 1798.

REMARKS *on a* SUBTERRANEAN WALL *in* NORTH-CAROLINA, *by the Rev.* ZECHARIAH LEWIS, *of the State of Connecticut.* [*See Medical Repository, vol.* ii. *p.* 272 *and* 275, 1*ſt edition.*]

SIR,

AGREEABLY to your requeſt, I tranſmit to you an account of the ſubterranean wall in North-Carolina.—This account was haſtily written, amidſt all the inconveniences of a journey. Since my return I have not found leiſure to reviſe and correct it; and, at preſent, can only tranſcribe it in its original crude form.

   I am, Sir,
    With ſentiments of eſteem,
     Your obedient friend and humble ſervant,
       ZECHARIAH LEWIS.

Dr. MILLER.

Weſt-Greenwich, Nov. 12, 1800.

THE celebrated ſubterranean wall, which, for ſeveral years paſt, has been made the ſubject of much converſation and inveſtigation in various parts of the United States, is about twelve miles north-eaſt of Saliſbury, in the county of Rowan, State of North-Carolina. It ſtands on uneven ground, near a ſmall brook. The ſtones of which it is formed are all of one kind, and evidently contain a quantity of iron ore. They are of various magnitudes; all of a long figure, and from one to twelve inches in length. The ends of the ſtones, which are of different figures, form the ſides of the wall. Some of the ends are ſquare; others are nearly of the form of a parallelogram, triangle, rhombus, or rhomboides; but moſt of them are irregular. Some preſerve the figure and dimenſion of the end through the whole length; others enlarge or diminiſh from the end. The ſurface of ſome is plain, of ſome concave, and of others convex. Every concave ſtone is furniſhed with one of a convex ſurface. Where the ſtones are not ſo exactly fitted as to lie perfectly level and firm, they are curiouſly wedged with others, which are very ſmall, and of a plano-convex form. The moſt irregular and unmanageable ſtones are thrown into the middle of the wall. The whole appear to be arranged in the moſt ſkilful manner to make the wall ſolid and ſtrong.

Every ſtone is covered with a ſpecies of cement. The cement, contiguous to the ſtone, has the appearance of iron ruſt; and, where it is thin, the ruſt has penetrated through it. Many pieces, however, are found more than an inch thick. In theſe the cement appears to be of a fine and curious texture; not the leaſt ſand nor grit is diſcoverable. In the wet parts of the wall, the middle of the cement, which is not diſcoloured with ruſt, is nearly of the colour, the conſiſtence, and the ſoft, oily feeling of putty in its ſofteſt ſtate.

The width or thickneſs of the wall is uniformly twenty-two inches: its height and length have not yet been diſcovered. The Rev. Dr. M'Korkle, Rev. Mr. Hall, and Dr. Newman, began, with conſiderable ſpirit, to give it a thorough examination. A ſum of money was collected, and a number of workmen employed, who devoted ſeveral days to the buſineſs. About forty yards from the part firſt diſcovered, and on a line with the direction of that part, they dug a large pit. After digging ſeveral feet they ſtruck the wall. The pit was then continued on one ſide ſome depth, that the baſe of the wall might be found, and its height aſcertained. The workmen, however, were ſoon incommoded with water, which roſe in the pit with great rapidity. At length, the man who was digging, although long experienced in the buſineſs of ſinking wells, became alarmed with his ſituation, and totally relinquiſhed the object. They then proceeded to diſcover the length. Another pit was dug beyond the former, and the wall again found. In the attempt to find the northern limit, a bend or off-ſet was diſcovered; after which the wall was found to proceed in a line parallel with its former direction. The off-ſet is not rectangular, but a gentle bend or curve.

The perpendicular diſtance of the parallel lines is about ſix feet. Without finding the terminations or the height, the fund was exhausted, the workmen fatigued and discouraged, and the investigation relinquished. The whole length yet discovered is about 300 feet, and the greatest height is twelve or fourteen. Both sides of the wall are plastered with a substance which resembles the cement on which the stones are laid.

The wall has been so broken by its numerous visitants, that the top of it cannot now be examined with accuracy. The Rev. Mr. Hall gives the following account of it: "Above the cavern or part first discovered, where the ground rises abruptly, the wall ascends; and below the cavern, where the ground gently declines, the top of the wall runs nearly parallel with the declivity. So far as the wall has yet been discovered, the highest part is fifteen or twenty feet above the lowest."

It appears to be the prevailing opinion, in this region, that the wall is a production of nature. In support of this opinion the following inquiries are suggested:

In the first place—What could have been the object of the wall? From the length already discovered, it cannot have been the part of any house or public building; and the ground on which it stands must have been badly calculated for a fortification.

In the second place—To build the wall with such stones—to fit them so exactly to each other—and, particularly, to collect, and so curiously arrange those of the smallest size, must have been the work of much time and immense industry. How is it supposable that men, in a rude state of society, would submit to the labour, or could possess the skill?

In the third place—Where could the materials have been procured?

In answer to the first of these inquiries, it may be observed, that, although we cannot, with absolute certainty, determine the object of the wall, yet we may form conjectures which contain, at least, as great probability as those which can be formed on the opposite side of the question. The form of the wall, so far as it has yet been discovered, will justify the supposition that it was built for a wall of defence. Its strength, also, at a time when bows and arrows were the principal instruments of war, would have been sufficient for the purpose. Since the wall was erected, the surface of the ground may have been greatly changed by repeated and heavy rains, and, possibly, by the general deluge.

It may have been built for the purpose of enclosing a prison, a garden, or a city.

It is also possible that the small stream which runs near the wall was formerly a considerable river, and that the wall was erected to guard against the rise of its waters.

To the second it may be replied, that we are not driven to the necessity of supposing the wall to have been built in a very rude state of society. It may have been constructed by some enlightened antediluvian nation, or by some civilized people who may have wandered to this region from the eastern continent. A tribe, not more refined than the native Mexicans, might have possessed sufficient skill. There are many productions of that nation, and even of more savage tribes, which exhibit more ingenuity and greater perseverance. The floating gardens, which furnished the Mexicans with subsistence; and the dyke, nine miles in length, which guarded them against the inundations of the lake, required incomparably greater efforts of industry, perseverance and skill.

To the third a satisfactory and full answer is at hand. On the adjacent hill are many stones evidently of the same substance and form: they are, however, larger and more broken than those in the wall. From this fact it appears that the best were selected for the work. The distance over which the stones were carried is not more than two or three hundred yards, and nearly the whole is descending ground. This is, possibly, the very reason why this spot was chosen for the location of the wall.

In a river, not more than one fourth of a mile distant, are found great quantities of muscle-shells, of which the cement may have been made. Mr. Probley, who lives within eighty yards of the wall, and who came to us while we were examining its materials and construction, assured us that he had carefully dried the cement, burnt the shells, compared the two, and could find no difference between them.

At a small distance there is also found a species of clay, resembling fuller's earth, of which the inhabitants make a convenient and durable cement for their chimneys. This cement is free from grit, and very similar to that of the wall.

The wall is perfectly regular, and has every possible appearance of an artificial production. The stones are all laid across the wall, and in a horizontal direction. The earth contiguous to it evidently appears to be what is generally termed made ground. It is of a grey, sandy character, and must have been washed to the wall by repeated and heavy rains.

Philosophers have generally adopted the opinion that there

are but three methods by which nature produces regular forms in the mineral fyftem. Thefe are, firft, that of cryftallization; fecondly, the crufting or fettling of the external furface of a liquid mafs while cooling; and, thirdly, the burfting of a moift fubftance while drying.* From a brief inveftigation, it will, I believe, be determined, that the wall cannot have been produced by either of thefe modes.

The different ftrata of fubftances formed by the firft mode are never found, in any confiderable quantities, to run in the fame direction, and never, I believe, horizontally. They generally run in oblique directions, inclining either to or from each other. But in the wall the different ftones are placed parallel with each other, and in a horizontal direction.

The fecond method is, the crufting of the external furface of a melted mafs. We cannot conceive it poffible that a mafs of melted fubftance, feveral hundred feet in length, twelve or fourteen in height, and precifely twenty-two inches thick, fhould ftand and cool in that pofition. The top would fink, and the fides fpread; and this, probably, not equally in all parts. But were it poffible that the liquid mafs fhould thus ftand and cool, it would be attended with the following procefs. The external parts firft lofing their heat, would firft congeal: the cold would contract the liquid ftratum, and thus feparate it from the remaining liquid fubftance. In the fame manner other ftrata would be formed, as the refrigeration fhould penetrate deeper into the mafs. Thus a ftratum would be formed on the top of the wall parallel with its length; and ftrata would, at the fame time, be formed on the fides, either parallel with its length, or perpendicular from the fummit to the bafe. As the middle of the mafs would not cool fo faft as the fides, fo the ftrata could not be formed horizontally from one fide to the other.

The third method, to wit, the burfting of a moift fubftance while drying, is fimilar, in its effects, to the fecond; of courfe fimilar remarks will apply. In this cafe the fubftance is in a moift inftead of a melted ftate. The burfting of the moift fubftance, or the feparating of the ftrata, is the effect of contraction by drying. The drying advances, in this cafe, juft in the fame manner with the refrigeration in the former; and, for the fame reafons, the wall cannot have been produced by the third method.

* Vide Encyclopædia.

Thus it is evident, in the first place, that no substantial argument is offered to prove that the wall cannot be artificial: in the second place, that it has every possible appearance of being the production of art: and, in the third place, that it cannot have been produced by nature in either of its ordinary modes of producing regular forms.

Essay on the FORMATION OF ROCKS, *or an Inquiry into the probable Origin of their present Form and Structure. By William Maclure.*

Our knowledge of the actual and present state of the substances which constitute our globe, is unfortunately confined to a small portion of the surface; from which it would appear, that we are still very deficient even as to those facts which are within the reach of our observation and experience, and which may perhaps be necessary to the forming of any rational conjecture concerning the formation or former state of those substances which cover the external surface of the globe.

Concerning the nature and properties of the great mass which constitutes the interior of the earth, we are entirely ignorant; few of our mines penetrate deeper than one fifty thousandth part of the earth's diameter under the surface, and none of them go beyond one twenty-five thousandth part of that diameter: it would appear, therefore, that any mere supposition concerning the actual and present state, or the nature of those substances which form the interior of the earth, is unsupported as yet by any reasonable analogy; and that all conjectures concerning former changes, partial or total, in the nature and structure of those substances, are removed still farther from any thing analogous in our present state of knowledge.

The earth being flattened at the poles, does not necessarily imply its former fluidity; we may be permitted to doubt the analogy between our experiments on bodies moving in our atmosphere, and the earth's motion in space: our total ignorance of the nature of the fluid which occupies what is usually called space, tends to render the analogy inconclusive.

May not the mode of casting patent shot be considered as an experiment on the form which liquid bodies would take by a rotatory motion? A drop of melted lead let fall from the height of two hundred feet is completely globular, and not flattened at the poles; the lead might be thrown with force from the top of the tower, which would imitate the centrifugal force, as gravitation does the centripetal force, and make the experiment more analogous.

The supposition that the earth was in a fluid state when it took its present form, leads to the supposition that it was always so; and that fluidity was the original state of the earth, kept so by all the general laws and order of nature, all of which general order and laws of nature must be totally changed before the earth would take a solid form.

On the supposition that the earth, previous to its fluid state, had existed always in a solid state, and that some creation or accident produced the fire or water necessary to its liquefaction, we have in that case first to suppose, that the order and nature of the general laws which had kept it always in a solid state, were totally changed, to produce a fluid state; and that another change in the general laws which produced and kept it in a fluid state, must have taken place previous to its having become again solid.

It may be doubted, whether the uniformity, order and regularity of the general laws of nature, which have at any time come within the limits of our observation, can warrant a supposition, founded on such complete changes in the mode of action.

The neptunists admit the fluidity of the earth, and endeavour to prove that water must have been the cause of that fluidity; though to *dissolve* the greater part of the substances now found on the surface, or as far under it as we have yet penetrated, would require two or three thousand times more water than the solid contents of the whole globe. How nature has disposed of that immense quantity of water, now become unnecessary by the consolidation of the globe, is but one of the many difficulties which arise out of the neptunian system.

The volcanists, likewise, consider the fluidity of the globe as a necessary foundation for their system; but insist that fire must have been the cause of it, nor can they, in a satisfactory manner, dispose of the immense quantity of heat or caloric, become unnecessary by the consolidation of the globe: difficulties that must always attend suppositions of a total change in the general laws of nature, because the agents necessary to the retaining of matter in one state, must be disposed of before that matter can acquire a different form or nature.

It is, perhaps, an historical fact, that all geologists who have formed their systems on the examination of the northern parts of the continent of Europe, where there are no existing volcanoes, are neptunists; and those who have examined Italy, or other volcanic countries, previous to the formation of their systems, are more or less volcanists, which tends to prove, that opinions are the result of our knowledge, and our knowledge the consequence of the different situations which chance or choice has thrown us into: we ought, therefore, not to be astonished, much less irritated, at the difference of opinions, but consider them as the natural effect or consequence of our locality or opportunities.

Suppose the earth was a body of moderate size, that we could cut up and dissect as we do animals, vegetables, or other objects of natural history, it is probable that the first part which would attract our attention would be the volcanoes, in action, with the mountains formed by the ejected matter; we should probably first examine the nature of this ejected matter, to ascertain what proportion of the surface of the globe, or ball, was covered with similar matter: we should, of course, find out the extinct volcanoes, and though the fire had ceased to act, the similarity and relative position of the matter would induce us to conclude that they were produced in the same manner, as well as the small detached remains of similar substances, which we would find scattered over the whole surface.

After shaving off all that we supposed to be formed by fire, the next active agent that would attract our attention would be water. The productions and changes wrought by the operation of this agent, would be examined: the aggregates of rounded particles, deposition with organic matter, &c. would be considered as belonging to formations by water.

Clearing the surface of the ball with our dissecting chisel of all that we could ascertain by analogy to belong to the formation by water, or fire, we would come to a species of matter that did not exactly resemble either of the above formations, which, on examining, we would find of various textures; and comparing it with the portions already cut off, we would find part of this matter which had a distinct resemblance to that formed by fire, and part to that formed by water, but so mixed and confused together as to prevent our forming any distinct conclusions. After turning the ball two or three times, we would naturally wish to know what constituted the interior or central part; for which purpose we would cut it in two, and expose the interior to our examination and analysis, as we had before examined the exterior; and if we should find that the interior was fluid, and like a soft boiled egg, and only the exterior was solid, we might follow the analogy of the egg a little farther, and deduce the probability, that at some former period the exterior crust had been fluid, and had since become solid, by some operation of nature analogous to something we had ourselves observed.

On the contrary, should the examination of the interior of the ball, prove, that it consisted of a variety of solid substances, farther and farther removed from any resemblance with those we had observed as formed on the surface by fire or water, we should probably conclude, that these agents were not necessarily instrumental in the formation of those substances; and that we were totally ignorant of the process which nature may have adopted to form those substances, and we should doubt whether those substances had not always existed in that state. Thus would the investigation be left, until farther dissections, and the analyses of similar constituted balls, had thrown more light on the subject of our inquiries.

In this manner the examination of the origin of the rocks that form the external crust of our globe ought, perhaps, to be conducted; beginning with those substances that have been formed under the immediate evidence of our senses, and completely within the limits of our observation, either by water or fire, and proceeding to others having a direct resemblance, in structure, component parts, or relative situation, or united by the chain of positive analogy, to the same mode of formation; evidently deriving their origin from the action of the same agents of water or fire, until we come to the last crust, beyond which we cannot penetrate; then we must drop the thread of positive analogy, and not being able to make a cut to the centre of the globe, be content with probable conjecture.

At this point, where positive analogy finishes, and probable conjecture begins, will be the natural line which will divide the rocks into *two classes;* the first class will contain all those whose origin, either by fire or water, has taken place under the evidence of our actual observation, or those that can be traced by positive analogy to the same origin. The *second class* comprising all those rocks which have no positive analogy with either, yet containing some parts which have a distant relation to both the modes of formation.

As nature does not advance by large leaps, but by small and regular steps, leaving no marks in the chain of gradation on which we can place the limits of our artificial division, the line of demarkation between the first and second classes will be doubtful; and the rocks approximating on both sides, will not be well determined. The line also must change with the progress of our knowledge and discoveries, and rocks placed in the second class now, because we have not found analogous rocks in the first class, may change their place by new discoveries, and pass from the second to the first class, or from what may be called the unknown into the known, whenever future experience and observation have thrown light on their origin.

There is no question here concerning the relative period in which the different formations by water or fire have originated. This is difficult to ascertain; and from the numberless derangements in the original order, liable to many exceptions, nor is the necessity of it evident in the inquiry concerning the origin. Nothing within our observation proves the priority of one mode of formation over the other, nor militates against the probability of one formation often alternating with another, and it is more than probable that the reason we have so few instances of such an alternation on record is because there is so small a proportion of the crust of our globe accurately examined.

In attempting to separate the rocks, whose origin comes within the sphere of our positive knowledge, or positive analogy, from those whose faint and distant resemblance leaves the nature of their origin to conjecture, I am convinced, that neither my experience, knowledge, nor industry, are adequate to the task of comparing their various differences and resemblances, so as to form an adequate conclusion; but the faults and imperfections in the execution will not, perhaps, injure the propriety of the arrangement or method; for it has always appeared to me necessary to fix some boundary between the knowledge of facts which must increase with our experience, and the field of conjecture which may, perhaps, on the contrary, diminish as our positive knowledge augments.

It is probable that nature has many more ways of effecting the changes, in the form of rocks, than we are acquainted with; and that she employs many agents, the nature and properties of which we are as yet totally ignorant of; nor is it improbable that she may form the same rock by two or more different agents. When we pretend to limit the operations of nature, to suit our contracted ideas, we most probably do her injustice. To proceed from the known, which we see daily forming, towards the unknown, through a chain of reasoning strictly analogous, is perhaps all that our present knowledge will permit us to do.

It is not intended to give a description of all the particular rocks that may constitute a formation, or be subordinate to it, many of them, such as the Topaz-rock, (which has only as yet been found in a bed, in clay slate, forty or fifty feet broad, and from two hundred to three hundred feet long) would tend to confuse: a general description of the formation, with a few observations, is all that I shall attempt.

SYNOPSIS OF THE ORIGIN OF ROCKS.

As we do not comprehend either the creation, or annihilation, of matter, by the origin of rocks we mean the last change which produced their present form, and the agents that nature employed to give them that form, or effectuate that change.

FIRST CLASS. Of Neptunian origin.

First Order. Formed by nature under our observation, visible, and resting on the evidence of our senses:

Sand beds,   Brown Coal,
Gravel beds,   Bog Iron ore,
Sea-Salt,   Calcareous Tuffa,
Sandstone,   Calcareous depositions,
Puddingstone,   Silex from Hot-springs, &c.

Second Order, resembling, in structure, position, or component parts, the first order, the evidence of their origin resting on direct and positive analogy:

Coal,   Graywacke & Graywacke slate,
Gypsum,   Transition Sandstone,
Chalk,   Transition Limestone,
Compact Limestone,   Transition Gypsum,
Sandstone,   Transition Clay Slate,
Puddingstone,   Anthracite,
Rock-Salt,   Siliceous Shist.
Old Red Sandstone,

SECOND CLASS. Volcanic origin.

First Order. Thrown out of active volcanoes, and resting on the evidence of our senses·

Compact Lava, Mud Lava,
Porous Lava, Obsidian or Volcanic Glass,
Porphyritic Lava, Pumice Stone,
Scoria, Cinders, &c.

Second Order. Resembling the first order in structure, position, and component parts, having the remains of craters, with currents of lava diverging from them: though the fire, which may have formed them, is now extinct; the evidence of their origin resting on direct and positive analogy:

Basalt, Pearlstone,
Trap formation, called by Werner the newest Flœts Trap formation, Porphyry attending the Trap as above,
 Clinkstone ditto, &c.
Pitchstone,

Third Order. Where the rocks resemble the second in texture and component parts, but where all the craters, cinders, scoriæ, and most of the porous rocks, have been washed away, leaving only the solid parts, such as

Basalt, Pitchstone,
Trap, called by Werner the newest Flœts Trap formation, Porphyry,
 Clinkstone, &c.

These rocks resemble the volcanic in relative position, covering indifferently all the other classes of rocks, and in detached pieces, without any extensive continuity or stratification, but divided by vertical fissures, the proof of their origin resting on a more distant analogy than order second.

THIRD CLASS. The origin doubtful, resembling a little the second order of the first and second classes, but the analogy neither direct nor positive, amounting only to probable conjecture.

First Order. Such rocks as probable conjecture would incline to place in the Neptunian origin:

Gneiss	Clay Slate,
Mica Slate,	Primitive Limestone.

Second Order. Such rocks as probable conjecture would incline to place in the volcanic origin:

Hornblende,	Sienite,
Porphyry,	Granite.
Greenstone,	

The origin of rocks may first be divided into the known and the unknown. The two first classes contain the known, and the third class the unknown. Farther observations may change their situation, and place a rock, which is now in the unknown class, in the known class, by which means the unknown class will diminish as our positive knowledge increases, and in proportion as the known class augments.

[*Editor's Note:* The remainder of this article (pp. 271–276, 285–310, and 327–345) includes detailed descriptions of rock types and has been omitted here.]

HINTS ON SOME OF THE OUTLINES OF GEOLOGICAL ARRANGEMENTS, WITH PARTICULAR REFERENCE TO THE SYSTEM OF WERNER

William Maclure

INTRODUCTORY REMARKS BY BENJAMIN SILLIMAN

SOME years since, during Mr. Maclure's geological survey of the United States, the editor had the pleasure of passing a few days, in company with that gentleman, in exploring the geology of the vicinity of New-Haven. Near that town, junctions, on an extensive scale, between widely different formations, are to be observed. A radius of ten miles, with New-Haven for a centre, will describe a circle within which the geological student may find (with the exception of formations, unquestionably volcanic) most of the important rocks of the globe, and a radius of even six or seven miles will include the greater number of these. At, and near the terminations of the primitive ranges, there are rocks which appear to have, in a high degree, the characters of the transition class. Among them is the beautiful green marble of the Milford Hills, seven miles from New-Haven. Mr. Maclure visited that district, and even suggested the first hint which afterward led to the discovery of the marble. Doubts being entertained concerning some of the geological relations of those rocks, a letter was addressed to Mr. Maclure (then in Philadelphia) on the subject. His answer is subjoined.

In giving it to the public, the editor takes a liberty which he hopes the respectable author will pardon, because his production, although evidently never intended for the public eye, contains statements and opinions of no small importance to the young geologist, especially of this country.

Geology, at the present day, means not a merely theoretical and usually a visionary and baseless speculation, concerning the origin of the globe; but, on the contrary, the *result of actual examination into the nature, structure, and arrangement of the materials of which it is composed.* It is therefore obvious, that the opinions of those men, who, with competent talent and science, have, with a direct reference to this subject, explored many countries, and visited different continents, are entitled to pre-eminent respect. SAUSSURE, by his scientific journeys among the Alps, (although a limited district has given deserved celebrity to his own name, and, if it were possible, has thrown an additional charm of attraction over those romantic and sublime regions. Dolomieu has made us familiar with the productions and phenomena of volcanoes, those awful and mysterious laboratories of subterranean fire. Humboldt has surveyed the sublimest peaks of both continents, and examined the structure of the globe amidst the valleys of Mexico and the snows of Chimborazo and Pinchinca; and Werner, with opportunities much more limited, (confined indeed to his native country, Saxony) but with astonishing sagacity and perseverance, deduced from what he saw, a classification of the rocks of our globe, which, although not perfect, has done immense service to the science of Geology. In this distinguished group (to which other important names might be added) Mr. Maclure has unquestionably a right to be placed. Few men have seen so much of the structure of our globe, and few have done so much with such small pretensions. His work on American Geology is noticed with becoming respect even in Edinburgh,* that focus of geological science. His opinions on some of the more obscure and doubtful parts of the Wernerian geology are worthy of peculiar consideration; for they are founded on a course of observations vastly more extensive than Werner ever had it in his power to make. The name of Werner will always be venerated as long as geological science shall be cultivated, for geology owes more to him than to any other man; but his pupils should not now demand that implicit and unqualified adoption of ALL his opinions, which will allow no other question to be raised, than what Werner taught or believed.

With these explanatory remarks, the following extract of Mr. Maclure's letter is now subjoined:

DEAR SIR,

Your letter of the 26th June came just as I was embarking for Europe. The information it requires concerning the primitive trap and flint slate, the transition and secondary rocks, &c. &c. is difficult to give without the aid of specimens, and frequently requires the examination of the relative position of the strata before any correct idea can be formed. I will, however, endeavour to give you the little my experience has brought me acquainted with.

Following the nomenclature of Werner, I have given a list of his rocks; but in describing them there are many of his names which I do not use; because I never met with them. Primitive trap is one instance—I do not use trap as a substantive, except in describing that kind of trap which Werner calls the newest flætz trap, the nearest to which is your trap,† which covers the oldest red sandstone.

The primitive flint slate is in the same predicament. I have always found it on the borders of the transition, between it and the secondary.

Primitive gypsum I have not found.

---

* Vide Edin. Review for Sept. 1818. p. 374.
† Referring to the ridges of Greenstone near New-Haven.

What Werner calls primitive trap may perhaps be compact hornblende, or perhaps the newest flætz trap, when it happens to cover the primitive; for, this species of trap, like the currents of lava, covers indiscriminately all classes of rocks, and is one reason why I consider it as the remains of ancient lava.

Transition trap is a rock that I have not met with, and may perhaps be a part of the flætz trap that happened to cover the transition, without any immediate connexion, but like a current of lava, overlying all the classes of rocks it meets with. This misapplication of names naturally arises from the system of neptunian origin, on which the nomenclature of Werner is founded.

Greywake and greywake slate are aggregates of rounded particles of rocks, evidently the detritus of more ancient formations, and differ from the aggregates of pudding and sandstone of the secondary class, in the following properties, viz.

The aggregates of transition are harder and much more compact, than the secondary; they are also cemented by argil, taking a slaty form.

This cement is in much greater quantity, in proportion to the particles cemented, and has the appearance as if the cement at the time of formation, had a consistence sufficient to prevent the particles from touching each other.

They have, in common with all the transition rocks, a regular and uniform dip from the horizon, from 10 to 40 degrees; and sometimes more. This is perhaps the strongest mark of distinction which separates them from the secondary, which are horizontal, or follow the inequalities of the surface on which they were deposited.

The transition are distinguished from the primitive in being aggregates of rounded particles, having little or no crystallization, and containing, or alternating with strata, which contain organic matter.

The oldest red sandstone, with all its accompanying strata, I should incline to put into the transition, as having many of the properties of that class, and occupying the same relative situation in the stratification of the globe. It is at a constant dip (although small) from the horizon; the cement is in greater quantities in proportion to the particles cemented than in any of the secondary aggregates, &c. &c.

The character of the secondary is a horizontal position, that perhaps does not admit of the same facility of examining the relative situation of its stratification. The compact limestone is, probably, with reason, considered as the lowest of the secondary formation, and always under the coal formation, but it appears to me that the secondary is deposited in basins alongside of one another, and that each basin has a different order of superposition, according to the nature of the agents employed in the deposition; that it is a partial, and by no means a general deposition. The secondary aggregates of sandstone and puddings have been evidently beds of sand or gravel, and of course, in that state would be called alluvial, but when cemented together by the infiltration of water, carrying along with it lime, iron, or any other body capable of agglutinating the particles together, become rocks, and may alternate in all proportions.

I am therefore inclined to think, that in geology the best mode for the greatest part of the secondary would be to give the relative position of the strata of each valley or basin; and I am rather of opinion that they would all differ from one another.

The French and English basin having chalk for the lowest stratum, which has occupied the geologists of both countries for these 10 or 15 years, is perhaps the best known; yet they do not know the relative position of the chalk and coals, because coals have not been found in the same basin with chalk; coals occupy basins filled with different kinds of rocks, and have no resemblance to the rocks found covering the chalk

A VIEW

Of the Theories which have been proposed, to explain the Origin of Meteoric Stones.

BY JEREMIAH DAY,

PROFESSOR OF MATHEMATICS AND NATURAL PHILOSOPHY IN YALE-COLLEGE.

FROM the earliest periods of history, there have been numerous reports of the falling of bodies from the heavens. But, till within a very few years, the subject has been considered, as belonging, rather to fiction and poetry, than to sober philosophy. Men of science, aware, it would seem, of the difficulty of assigning a plausible reason for the descent of masses of stone from the atmosphere, have chosen to intimate, by a significant silence, their disbelief of the accounts which had been given of their fall. But the subject has lately intruded itself on their notice, in a way that has left them without an apology for refusing, any longer, to take it into consideration. A body of evidence has been accumulated, which it would require more ingenuity to explain away, than to account for the phenomena. They have accordingly come forward, at last, with a very liberal supply of hypotheses. These have been proposed, by different persons, and on various occasions; and are scattered among the transactions of philosophical societies, periodical publications, and fugitive papers. It may be of some service, to collect them into one view; and to compare them with each other, and with the facts which they are intended to explain.

Before a direct examination of the merits of these theories, it will be important to recollect, that the falling of stones is frequently, if not invariably, connected with another phenomenon, the passage through the air of one of those large luminous meteors, which occasionally make their appearance in our atmosphere. When the stones have fallen in the *day* time, the meteor has not always been observed: probably because its light was not sufficiently strong to draw the attention of persons abroad, to that part of the heavens in which it was moving. But, even in this case, the same kind of *report* has been heard, as that which usually follows the explosion of a meteor. In many instances, the luminous body has been seen to come forward to the zenith, and apparently to burst; and, immediately after, the stones have fallen, with a whizzing noise, to the ground.

Meteors of this kind are seen, in some parts of the world, almost every year. They appear moving through the heavens, like balls of fire, or red hot iron. Their apparent diameter is sometimes as large as the moon. From the main body, frequently extends a flame or train. Streams and sparkles of fire seem to shoot out, on every side. Just before their disappearance, there is a violent explosion; by which pieces often appear to be detached, and thrown to the ground.

The same meteor is seen over a great extent of country; in some instances, a hundred miles in breadth, and five hundred in length. Bodies seen from such distant places, at the same time, must have a great elevation. From various calculations, it appears, that, during the time in which they are visible, their perpendicular altitude is generally from twenty to a hundred miles. Their diameter is, in some cases, estimated to be at least half a mile.

Their velocity is astonishingly great. Though they are rarely visible, for more than a minute; yet they are seen to traverse many degrees in the heavens. Their rate of motion cannot, according to calculation, be generally less than three hundred miles in a minute.

It is not, in every case, known that these bodies project any thing to the earth. It is probable, however, that stones do, in many instances, fall from them without being noticed. For, as they bury themselves a considerable depth in the ground; there could be little chance of their being discovered, unless they were either seen or heard to fall. But the instances in which they have been actually observed to descend through the air, immediately after the explosion, are sufficiently numerous to establish the point, that the stones proceed from the meteor. The two phenomena, therefore, are really but one event: and no hypothesis can be admitted as a satisfactory explanation of the one, which does not, at the same time, account for the other. Whether any of the various conjectures which have been proposed for this purpose are founded in truth, must be determined by making a comparison between the leading features of each, and the facts to which they are intended to be applied.

1. One hypothesis is this; that the materials of which the meteoric stones are composed, are raised into the air in the state of exhalations or gases—that, in the upper regions of the atmosphere, they are occasionally collected in great abundance—that, some of them being inflammable, a combustion takes place—and, that the particles of the whole, by their mutual attractions, rush together, and form a mass, which descends by its weight, to the ground.

To this supposition there are several important objections. In the first place, the principal substances of which these bodies are composed, are never known to be raised in vapour. The ingredients are iron, silex, magnesia, nickel and sulphur. Several of these cannot be evaporated, even by the powerful heat of a furnace. In what way then can they be carried up, fifty or a hundred miles, from the surface of the earth?

But, supposing the materials, by some unknown process, and contrary to all our experience, to be carried into the air, and the bodies to be formed there; what is there in the atmosphere, which could give them their rapid horizontal velocity? A solid substance elevated to a great height, and left to itself, would *descend* very rapidly. But the motion of a meteor is not, like that of a falling body, perpendicular to the horizon, but almost parallel. Its velocity is such as could not be produced by the atmosphere. The air will not communicate, to a body floating in it, a motion more rapid than its own. The progress of the most violent wind is not more than two or three miles in a minute. But a meteor moves several hundred. The velocity of sound is less than twelve hundred feet in a second: that of a meteor, more than twenty thousand. The greatest force of gunpowder, will throw a cannon ball but a very few miles. A meteor is often seen to move several hundred. Is it not incredible, that a power sufficient to produce such a motion should reside in the atmosphere?

There is still another very weighty objection, to the supposition that these substances are formed in the air. It cannot be true, as the theory would imply, that the body of the meteor falls to the ground. The pieces which come down are only fragments, detached from a much larger mass. This is evident, from the size of the meteors. They are calculated to be several hundred feet in diameter. One which was seen in England and France in 1783, was computed to be almost two miles in circumference. From the various accounts which have been given of that which lately exploded over Fairfield County, in this state, it is evident, that it must have been many thousand times larger than the amount of all the stones which have been found. The whole that has been collected, would not form a sphere two feet in diameter. This, at the distance of fifty miles, would subtend an angle of less than two seconds of a degree: and, therefore, if seen at all, would appear like a fixed star, a mere visible point, too small to be measured by the nicest instruments. But the meteor from which these stones proceeded, had a very considerable apparent diameter, to those who saw it even at greater distances.* It is not pretended that the dimensions can be ascertain-

* NOTE....It was seen at Wenham, in Massachusetts, 150 miles from the nearest part of its path.

ed with any great degree of accuracy. The appearance is so sudden and unexpected, that no opportunity is afforded of measuring the diameter with an instrument. Some exaggeration is to be expected, from the novelty and splendour of the object, and the surprize of the observer. And it is possible that the apparent dimensions of the meteor may be enlarged, by the flame or glare of light, with which it is sometimes accompanied. But, after making ample allowances on all these accounts, the body will still remain vastly larger than any which has been known to fall to the earth. It cannot be the fact that the luminous object is *principally* flame or vapour. If this were the case, it could not preserve a regular globular figure while moving through the atmosphere with a velocity twenty times as great as that of sound. It would be immediately dissipated. The great body of the meteor must be a solid compact substance, capable of sustaining the resistance of the air. Its magnitude is such, as to illuminate, at once, a region of one or two hundred miles in extent. It is inconceivable, that a body only two or three feet in diameter, however luminous, should attract, at the same moment, the gaze of a whole country; and appear, to the distant spectators, one third, one half, or three fourths as large as the moon. The real diameter of the meteor, according to the lowest computation, must be some hundreds of feet. No such body has ever come to the ground.

Of the meteor which was lately seen in this state, the part that has gone from us, is many thousands of times as great as the sum of all the pieces which have been discovered. After throwing off a few hundred weight from its surface, it must have held on its course, and either revolved round the earth, or gone off to the distant regions of the heavens. It must require a strong faith to believe that the atmosphere, even if furnished with materials, could produce *such* a body, and then give it a velocity sufficient to carry it beyond the circumference of the earth.

2. A second hypothesis is, that the meteoric stones are masses of matter thrown from volcanoes. But this is embarrassed with difficulties as great as the one which

has already been examined. In the first place, the substances which are known to be thrown from volcanoes, are all of a different kind from these stones. No such bodies are found in their craters, or in the neighborhood where the lava has fallen. It may therefore be concluded that none such have ever proceeded from them.

In the next place, the stones have fallen at the distance of several hundreds or thousands of miles from any known volcano. It is next to impossible that they should be carried thus far, by any force which could be applied to them near the surface of the earth. The resistance of the air is so great, that it will not suffer the motion of a body in the lower regions of the atmosphere, to exceed a certain limited rate. This has been ascertained by the numerous experiments which have been made, for the purpose of improving the theory of gunnery. It is found that if the greatest possible velocity be given to a cannon ball, the air will almost immediately reduce it to about eleven or twelve hundred feet in a second. A larger body would indeed be retarded in a less degree. Still, the resistance would be such as to bring it soon to the ground. It requires an initial velocity, greater than that of sound, to carry a cannon ball only three or four miles. Is it not then incredible, that a body a quarter of a mile in diameter, should be thrown from a volcano, with a force sufficient to carry it hundreds or thousands of miles; and that, after having gone this distance through the atmosphere, it should still retain a velocity greater than that, with which a shot issues from the mouth of a cannon? But what decides the point, is, that the meteor after all, does not fall to the ground. It moves on, in a curve, which could not be described by a body, that had commenced its motion at the surface of the earth.

Mr. King has varied this theory,* to accommodate it to the particular circumstances of the fall of stones at Sienna in Tuscany. He supposes that these substances were thrown from a volcano, not in *solid masses*, but in the state of *dust* or *ashes*. Sienna is about two hundred miles from Vesuvius. The shower of stones was the

* See " Remarks concerning Stones said to have fallen from the clouds. By Edward King, Esq. F. R. S. London, 1796."

next day after a great eruption from that well known volcano. Though it is scarcely credible that *solid bodies* could be projected to this distance; yet it is thought not to be impossible that *pulverized* substances should be wafted thus far in the air. Mr. King supposes that a vast quantity of ashes, composed of particles of iron, sulphur, and other ingredients, was thrown from Vesuvius to a prodigious height—that they there formed a cloud, and floated to the region of the atmosphere over Sienna—that when they began to descend, they became so much condensed, as to take fire, and produce numerous explosions—that the pyritical, metallic and argillaceous particles were melted—and that when cooled again, they were consolidated into the masses which were seen to fall to the ground. It is unnecessary to enter into a minute examination of this theory; as it is framed to suit the peculiar circumstances of the fall of stones at Sienna, and is applicable to no other case. It may be sufficient to observe, that, excepting merely the difficulty of getting the materials into the air, it is liable to all the objections which have been stated to the supposition, that the meteoric bodies are produced by *exhalations* from the surface of the earth.

3. Some philosophers, dissatisfied with the methods which have now been mentioned, of accounting for the falling of stones from the heavens, have ascribed to them an origin still more extraordinary. They suppose them to be sent to us from the moon. By the aid of the telescope, luminous spots have been discovered on the moon, which, from their changeable appearance, are supposed to be volcanoes. If bodies can be projected from these, a certain distance towards us; they will never return, but will be drawn away by the superior attraction of the earth. There is a particular point, between us and the moon, in which, if a body were placed, the attraction of the earth and of the moon upon it would be equal. This point is calculated to be twenty-four thousand miles from the moon's center—about one tenth of her distance from the earth.

The velocity with which a body must be thrown from a lunar volcano, to reach this point of equal attraction, is about ninety miles in a minute—not more than one third of the velocity with which a meteor moves, when near the earth; and only three or four times as great as that with which a ball may be sent from the mouth of a cannon.

The atmosphere of the moon would probably oppose some resistance to the motion of a body passing through it; but far less than ours. It has so little extent and density, that its very existence has long been a subject of dispute among astronomers.

If the earth and the moon were at *rest*; and a body were sent directly from one to the other; it would strike it. But the moon has a revolution round the earth. Every body thrown from her surface, must partake of her motion in this orbit. The path described by a body projected from a lunar volcano, would not, therefore, be a *right line*, directed to the center of the earth; but a *curve*, which would be the result of a combination of the projectile force, the motion of the moon in her orbit, and the power of gravitation.* The body, instead of striking the earth, would probably revolve round it. In some part of its revolution, it might fall within the *atmosphere;* and pieces detached from it, by violence, might be thrown to the ground.

* Note....The editor of the late abridgement of the London Philosophical Transactions, (vol. vi. p. 108,) supposes, "that the apparent motion of the meteor, in a direction almost parallel to the horizon, may be owing to the motion of the earth, in its annual orbit: That, while the body is coming towards us, the earth *glides away*, and leaves it behind." If this were the fact, the motion of meteors ought always to be in a direction contrary to that of the earth in its orbit; and therefore in the plane of the ecliptic. But they are observed to come from various points of the compass. That which exploded over Weston, moved in a direction almost *perpendicular* to the ecliptic. Besides, the motion of the earth in its orbit, is not in a line parallel to the horizon, except at noon and midnight. And indeed, it is not easy to see, how the apparent motion of a body projected from the moon, could be *in any way* affected by the annual revolution of the earth. For the moon has the *same* revolution. A body thrown from one to the other, partakes of the motion, which is common to them both. A ball fired from a ship under sail, at another ship, moving in the same direction, and with the same velocity, would not be prevented, by the motion of either vessel, from striking its object.

This theory can perhaps claim one advantage over the others which have been mentioned, the merit of bare possibility. But its advocates appear not to have taken into consideration the *size* of the meteors, from which the falling stones proceed. They attempt to give an account of those small pieces only, which are actually found on the ground. But these must be a very small portion indeed, of a body a mile or two in circumference. According to the observations of Dr. Herschell, the altitude of the mountains in the moon, does not generally exceed half a mile. These can be little larger than some of the meteors which appear in our atmosphere. Before we acquiesce in this theory then, we must be prepared to believe, not merely that the lunar volcanoes throw a few pieces of lava, as far as the earth; but, that they send us whole mountains.

Nor is this all. Among a number of bodies, thrown at random from the moon, it is not probable, that one in ten thousand would have precisely that direction, and that rate of motion, which would be requisite to cause it to pass through our atmosphere, without falling to the ground. Yet a meteor is seen, in some part of the world, almost every year. To account for this fact, by the theory in question, we must suppose, that thousands of bodies are annually thrown from the moon, each of which is several hundred feet in diameter. A pile of mountains as large as the Andes, would, at this rate, be very soon scattered.

4. There is one other hypothesis, which, though not entirely without difficulties, appears to be encumbered with fewer than any other, which has been offered to the public. Among the manuscripts of the Rev. Thomas Clap, formerly President of Yale College, was found a paper, containing " Conjectures on the nature and motion of Meteors." This was published, some years after his death. It is thought, that the theory of " Terrestrial Comets," which it proposes, may be so modified, as to suit the case of atmospheric stones.

The solar comets, it is well known, revolve round the sun, in very eccentric orbits. In one part of their revolution, they sometimes come so near as almost to strike his body. They then move off, far beyond the orbits of all the planets; and, in some instances, are gone hundreds of years, before they return. President Clap supposes, that the earth is furnished with its system of comets, as well as the sun----that their size, and the period of their revolutions, are proportioned to the comparative smallness of the primary body, about which they revolve----that, like the solar comet, they fly off, in very elliptical orbits; and, during the greatest part of their circuit, are too far distant to be visible----that, in their approach to the earth, they fall within our atmosphere----that, by the friction of the air, they are heated, and highly electrified----that the electricity is discharged with a very violent report----that they then move off in their orbits, and, by their great velocity, are soon carried out of our sight.

It does not appear, that the learned author of this theory was apprized of the fact, that substances frequently fall from these bodies to the ground. But the scheme requires very little alteration, to accommodate it to this circumstance. We have only to suppose, that, at the time of the explosion, pieces are broken off from the surface of the meteor; and that these fall to the earth, while the main body moves on in its orbit.

The hypothesis, if admitted, will account for most of the phenomena attending the fall of atmospheric stones. The *velocity* of the meteor corresponds with the motion of a terrestrial comet, passing through the atmosphere in an elliptical orbit. A body moving near the earth, with a velocicity less than 300 miles in a minute, must fall to its surface by the power of gravitation. If it move in a direction parallel to the horizon, more than 430 miles in a minute, it will fly off in the curve of an hyperbola; and will never return, unless disturbed in its motion by some other body besides the earth. Within these two limits of 300 miles on the one hand, and of 430 on the other, (some allowance being made for the resistance of the air, and the motion of the earth,) the body will revolve in an ellipsis, returning in regular periods. Now, the velocity of the meteors, which have been observed, has generally been estimated to be rather more than 300 miles in a minute. In some instances, it is perhaps too great, to suffer the body ever to return. But, in most cases, it is calculated to be such as would be necessary, in describing the lower part of an elliptical orbit.

The *direction* of the motion also, agrees with that of a revolving body; but not at all with that of a mass of matter, accumulated in the atmosphere, and falling, by its weight, to the earth. The *dimensions* of these meteors too, are such, as to indicate, that they move in orbits of their own; as they are manifestly too large to be formed in the air, by an accumulation of gases, or to be thrown from a volcano or the moon. They appear to have about the same proportion to their central body, the earth, as the little planets lately discovered between the orbits of Mars and Jupiter, have, to the sun, about which they revolve.

The theory last stated, though in the main adapted to the purposes for which it was proposed; yet, it must be acknowledged, is not entirely satisfactory, in the explanation of one or two particulars. It assigns a reason for the ignition and explosion of the meteor, which is not perhaps fully warranted by any observations and experiments hitherto made. The stones, when they fall to the ground, are found to be hot. The body of the meteor itself, has the appearance of fire. It is undoubtedly in a state of ignition, at least at the surface. Whence is this powerful heat derived? President Clap supposes it is produced by the friction of the air----that the body, moving with great rapidity through the atmosphere, is both heated and electrified----and that, when it is nearest the earth, the electricity is discharged, with an explosion, as much greater than thunder, as the meteor is farther distant, than the common region of the clouds. It is well known, that *hard* substances may be electrified, and even set on fire, by rubbing them together. But farther proof is wanted, to make it evident, that a body may be made red hot, by the mere friction of the air; especially of air, as greatly rarefied, as it must be, in that part of the atmosphere where the meteors move.*

*Note....Since the discovery of Mr. Davy, that the earths are metallic oxides; it has been suggested, that the bases of magnesia and silex, may originally exist in the meteor, in the state of *pure metal*: and that, when the body comes from some distant region of the heavens, into our atmosphere, a sudden and violent combustion is produced, by the very strong affinity of these substances to oxygen.

There is another circumstance, which is left unexplained, by this theory. In a few instances, particularly that at Sienna, the falling of stones is said to have been accompanied, or preceded, by an apparent burning of the clouds. If this is any thing more than an optical deception, it seems to indicate, a collection of combustible materials in the air. This appearance of fire in the heavens, has been too long before the falling of the stones, to be the *effect* of the passage of the meteor through the atmosphere.

With the exception of these two difficulties, neither of which ought perhaps to be considered as insuperable, the theory, which refers the origin of the meteoric stones to terrestrial comets, appears to be embarrassed with fewer objections, than any of the others which have now been mentioned. None of them, however, can claim to be considered as any thing *more* than theories. They are not yet supported by direct and positive proof. The subject is involved in too much obscurity, to admit of a complete elucidation at once. The enquiry has commenced, with a number of suggestions, which *may* be true; but which must be left, to be confirmed or refuted, by subsequent observations. This is, not unfrequently, the course which scientific investigations must of necessity take. The first step towards an important discovery, is often an ingenious conjecture. This gives the lead to a train of inquiries, which finally succeed, in unfolding the true principles of the subject. It must be granted, that but little progress has, as yet, been made, in explaining the origin, nature, and use, of the bodies, from which the atmospheric stones proceed. But the facts that have been collected, have awakened curiosity. The approach of these meteors, will hereafter be noticed, with uncommon interest. Observations of their motion, will probably be made, with as much accuracy, as the opportunities furnished, by their sudden and unexpected appearance, will admit. But whether the mysteries of the subject will be unveiled, upon a farther investigation, time must determine.

Part VII

FIELD GEOLOGY

Editor's Comments on Papers 23 Through 28

23 **MITCHILL**
   Geological Remarks on Certain Maritime Parts of the State of New York

24 **MACLURE**
   Observations on the Geology of the United States, Explanatory of a Geological Map

25 **MITCHILL**
   An Amendment Proposed to the Geological Chart of the United States, as Respects the Character of the North Side of Long Island, Which Is Shown to Be Alluvial and Not Primitive, as Therein Stated

26 **RAFINESQUE**
   Review of Observations on the Geology of the United States of America; With Some Remarks on the Effects Produced on the Nature and Fertility of Soils, by the Decomposition of the Different Classes of Rocks, and an Application to the Fertility of Every State in the Union, in Reference to the Accompanying Geological Map—With Two Plates, by William Maclure

27 **RAFINESQUE**
   Review of An Index to the Geology of the Northern States, with a Transverse Section from the Catskill Mountains to the Atlantic, Prepared for the Geological Classes at William's College, Massachusetts, by Amos Eaton

28 **HAYDEN**
   Agenda, or Selection of Queries

Field investigation of the areal distribution of rocks and minerals provided at once the greatest challenge and the greatest potential benefits for the young United States. The nation relied on natural resources for many daily needs and looked to these resources for growth and development. In spite of the hardships and privations of travel in the interior parts of North America, several workers prior to 1820 delineated important aspects of the continent's geological structure.

The earliest observations on North American geology were made by the continent's first explorers (White 1969). The search for gold and other natural treasures was a motivating factor in early exploration so the minerals and rocks of the uncharted lands were of prime interest. The many isolated observations of natural productions in the New World enabled Jean Étienne Guettard (1715-1786) to compile a memoir on the minerals and rocks of North America with a mineralogical map. White (1969) also credited Lewis Evans (1700-1756), Peter Kalm (1716-1779), Major Robert Rogers (1731-1795), Jonathan Carver (1732-1780), and Thomas Hutchinson (1730-1789) with important contributions to the acquisition and publication of data on America's geological structure.

Johann David Schöpf (1752-1800) is usually credited with having completed the first systematic study of North American field geology, and his work has received considerable analysis (Schöpf and Spieker 1969). Schöpf's efforts did not reach a contemporary American readership, however, and consequently had little influence on the development of the earth sciences in the United States (Spieker and Hazen 1975).

Contrary to common belief, an American worker had also engaged in systematic field observations by 1787, for it was in that year that Samuel Latham Mitchill published his little-known "Geological remarks on certain maritime parts of the State of New York." The "Geological remarks" were originally an addendum to a biological pamphlet on "The absorbent tubes of animals" but were republished as a separate article in the *American Museum or Universal Magazine* of 1789 (Paper 23). Mitchill, noted previously for his work in medical geography, had a compulsively inquisitive mind and engaged in the study of New York rocks "to increase [his] knowledge of physics, and to become a practical naturalist, and [he] on this account, walked over a considerable tract of land."

Mitchill's brief account is noteworthy for several reasons. It contains what is probably the earliest use in an American publication of *geology*, a word not generally accepted in Europe until after 1778 (Adams 1932). Regarding the geology of the area near New York, Mitchill stated that it is primarily granite and therefore primitive, rather than secondary. In his description of local fossils (in this context all rocks, minerals, and organic remains), Mitchill listed quartz, mica feldspar, and shoerl (probably both hornblende and tourmaline) as the granite's principal constituents. These observations were used to question the generalization that all coastal areas of eastern North America were former sea beds, now secondary or alluvial sediments. Mitchill clearly recognized the powers of gradual erosion to shape a landscape, for he suggested that New York Island (i.e., Manhattan) was once connected to the adjacent parts of Long Island and to other local granitic outcrops.

The most influential and comprehensive eighteenth-century account of North American geology was *A View of the Soil and Climate of the United States* by the Frenchman Constantine François Chasseboeuf, Comte de Volney (1757-1820). Based on extensive travels and research from 1795 to 1798, Volney's text reached a widespread audience with French, British, and American editions available by 1804. The numerous observations on North American rocks, minerals, and geologic structure contained in this account have been analyzed in detail by White (1968) in a modern reprint edition of the text. The significance of Volney's work lies, in part, in the increasing recognition of the geological structure of eastern North America, as demonstrated by his geological map. He collected a large amount of field data and reduced these to a more systematic framework, which allowed his readers to comprehend the broad outlines of rock distribution in the United States. Volney's data, along with those of Mitchill and others, were further condensed in James Mease's (1807) *Geological Account of the United States,* which was the first publication on American geology directed toward a popular readership.

The most influential contribution to American geology before 1820, and certainly the grandest in scope, was William Maclure's "Observations on the geology of the United States." This classic work appeared in the 1809 *Transactions of the American Philosophical Society* (Paper 24). Though the text only included 18 pages, the accompanying geological map represented field investigations of over half a million square miles. Maclure is reputed to have journeyed across the Allegheny Mountains more than fifty times on horseback while compiling his memoir and map. The 1809 "Observations" commenced with a brief explanation of his preferred rock-classification system. Wernerian nomenclature was adopted because it was "the most perfect and extensive in its general outlines" and because Werner's concept of a globally constant sequence of rock types *might* be found to apply to American formations. Maclure was ready to accept Werner's theory only if the field evidence supported it.

Maclure's essay cataloged locations of the primitive, transition, secondary, and alluvial rocks of America (see Part 6). The areal extent of these formations was illustrated on his geological map, which bears a striking resemblance to modern geological maps of eastern North America. The article closed with the modest hope that others "possessed of more talents and industry might continue the work."

Maclure's study provided an excellent data base on which local geologists could expand. Virtually every American field geology publication in the following twenty years made some reference to the "Observations." A few of these articles, such as S. L. Mitchill's "Amendment proposed to the geological chart of the United States" (Paper 25), were specifically designed to correct errors in Maclure's map. Mitchill noted Maclure's mistake in supposing that the northern side of Long Island, New York, was composed of crystalline rocks. A tabulation of alluvial deposits for several locations was given to support the opinion that virtually all of Long Island is composed of unconsolidated sediments.

Mitchill's observations, along with many other additions and corrections, were incorporated in the 1817 second edition of Maclure's "Observations," a pamphlet of 128 pages with map and cross sections.* Maclure in his 1817 "Observations" used the same Wernerian rock classification presented in the 1809 version, though he was careful to note that the Wernerian method was "subject to all the errors inseparable from systems founded upon a speculative origin." The first two chapters are in other respects also similar to the earlier edition, but two additional chapters on the decomposition of rocks and the consequent effects on soil fertility are original to the 1817 pamphlet.

This expanded version was met with universal acclaim (Papers 26 and 29). The review by Constantine Samuel Rafinesque (1783-1840) is particularly interesting, for Rafinesque was himself a distinguished geologist and naturalist of the American Midwest. The reviewer praised Maclure for his "zeal, assiduity, perspicuity, liberality, utility, and an early attention to [geology]." Rafinesque continued: "More essential facts and truths are disclosed than in many thick volumes of yore." In spite of this praise, Rafinesque realized that Maclure's study was only the beginning. "We cannot but regard his efforts, as well as those of Volney, as mere attempts towards the knowledge which he means to convey. . . . A long period must elapse before we can acquire a complete knowledge of the soil we inhabit." Maclure laid a sure foundation on which the more detailed studies of future decades were constructed.

Numerous field geology reports appeared shortly after the 1817 "Observations." Geological maps of Boston (Dana and Dana 1818), central Massachusetts (Hitchcock 1818), and the region south of Albany, New York (Eaton 1818), closely followed Maclure's publication and introduced for the first time three of the most eminent names in American earth science. Boston chemists James and Samuel Dana were, respectively, uncle and father to James Dwight Dana (1813-1895), the most distinguished of all American mineralogists. Edward

---

*The 1817 edition was also published in the 1818 *Transactions of the American Philosophical Society* and has recently been reprinted by the Society.

Hitchcock (1793-1864) was later to complete a comprehensive survey of Massachusetts geology and was professor of geology and subsequently president of Amherst College. Amos Eaton (1776-1842), like Hitchcock, distinguished himself in both geology and education. Eaton spent his early professional years as a lawyer but demonstrated his interests in natural history by publishing an elementary botany textbook that long remained the standard American work. Eaton's geological career did not commence until he reached the age of 40, at which time he produced *An Index to the Geology of the Northern States*, in addition to the geological map mentioned previously.

Eaton's *Index*, though less ambitious in scope than Maclure's "Observations", was nonetheless a substantial undertaking. After cataloging the geographical distribution of five main formations in a region including much of New England and eastern New York, Eaton proposed a generalized geological cross-section for the northern states (see Merrill 1924, p. 57). This effort was favorably reviewed by Rafinesque (Paper 27) shortly after his review of Maclure's great study. Rafinesque recognized that Eaton's account presented a more satisfactory picture of the geological structure of this region than the "Observations," which portrayed the entire area as undifferentiated primitive rocks. Even so, "many other perambulations and excursions, and much research, are yet requisite, before a complete idea of the soil . . . could be formed." Rafinesque favored Eaton's cross-sections over those of Maclure but argued that the former were projected to unreasonable depths on the basis of contemporary knowledge. In spite of such minor criticisms, the efforts of Eaton and his fellow field geologists were eagerly anticipated and well received.

The methodology of field geology was well established by 1820. Workable rock and mineral classifications were generally available in the publications of Cleaveland and Maclure, and a variety of theoretical frameworks existed for arranging the field data. This advanced state of the earth sciences is reflected in the "Geological Queries" (Paper 28), published as an addendum to Horace H. Hayden's *Geological Essays; or an Enquiry Into Some of the Phenomena to Be Found in Various Parts of North America and Elsewhere*. Hayden (1769-1844) was a resident of Baltimore and an active investigator of local geology, although his chief occupation was as a successful dentist. His essays were an attempt to relate the coastal plains of the Atlantic seaboard to the Noachian deluge, rather than more gradual mechanisms of deposition. Hayden's theory was not accepted by most geologists, though it was reprinted in theological journals and used in support of a literal interpretation of the Bible. The "Geological Queries" added to the *Essays* revealed no particular prejudice regarding the origin of rocks and soils but nevertheless indicated the advanced state of field geology in 1820.

The queries provided a guide to the systematic field examination of an area. The researcher was advised to examine such broad topographical features as mountains, valleys, and rivers, with a view toward the elucidation of both the underlying geological structure and local rocks, minerals, and organic remains. If the ability to successfully classify and organize data into a logical framework is a measure of scientific progress, then geology had been founded in America by 1820.

REFERENCES

Adams, F. D. 1932. Earliest Use of the Term Geology. *Geol. Soc. America Bull.* **43**: 121-123.

Dana, J. F., and S. L. Dana. 1818. *Outlines of the Mineralogy and Geology of Boston and Its Vicinity with a Geological Map.* Boston: Cummings and Hilliard.

Eaton, A. 1818. *Index to the Geology of the Northern States.* Leicester, Mass.: H. Brown.

Hitchcock, E. 1818. Remarks on the Geology and Mineralogy of a Section of Massachusetts on the Connecticut River, with a Part of New Hampshire and Vermont. *Am. Jour. Sci.* **1**:105-116, 436-439.

Merrill, G. P. 1924. *The First One Hundred Years of American Geology.* New Haven, Conn.: Yale Univ. Press.

Spieker, E. M. 1972. *Geology of Eastern North America by Johann David Schöpf. An Annotated Translation.* . . . New York: Hafner.

Spieker, E. M., and R. M. Hazen. 1975. The Founding of Geology in America, 1771 to 1818: Discussion and Reply. *Geol. Soc. America Bull.* **86**:1615-1616.

White, G. W. 1968. Introduction to a reprint edition of C. F. Volney's *View of the Soil and Climate of the United States.* New York: Hafner.

White, G. W. 1969. Early Geological Observations in the American Midwest. In *Toward a History of Geology*, edited by C. J. Schneer. Cambridge, Mass.: MIT Press, pp. 415-425.

BIBLIOGRAPHY

Akerly, S. 1810. A Geological Account of Dutchess Co. in New York. *Am. Mineralog. Jour.* **1**:11-16.

Akerly, S. 1820. *An Essay on the Geology of the Hudson River and the Adjacent Regions.* New York: A. T. Goodrich, 69p., plate.

Atwater, Caleb. 1819. Notice of the Scenery, Geology, Mineralogy, Botany, etc. of Belmont County, Ohio. *Am. Jour. Sci.* **1**:226-230.

Buel, David, Jr. 1819. Observations on the Geology of the Counties of Montgomery and Schenectady, in the State of New-York. *Plough Boy* **1**:232.

Columbian Magazine or Monthly Miscellany. 1786. Some Observations on the Structure of the Surface of the Earth in Pennsylvania and the Adjoining Countries, vol. 1, pp. 49-53.

Dewey, C. 1819. Sketch of the Mineralogy and Geology of the Vicinity of Williams College, Williamstown, Mass. *Am. Jour. Sci.* **1**:337-346.

Dewey, C. 1820. Geological Section from Taconick Range, in Williamstown, to the City of Troy, on the Hudson. *Am. Jour. Sci.* **2**:246-249.

Drake, D. 1810-1811. Strictures on Volney's "View of the Soil and Climate of the United States." *Port Folio* **4** (third series):587-591; 320-324; **6** (third series): 203-209.

Eaton, A. 1820. *Index to the Geology of the Northern States*, 2d ed. Troy, N.Y.: W. S. Parker and Albany, Websters and Skinners, xi, (1), (13)-286p., 2 plates.

Eaton, A., and T. R. Beck. 1820. *A Geological Survey of the County of Albany.* Albany, N.Y.: 56p., table.

Evans, Lewis. 1755. *Geographical, Historical, Political and Mechanical Essays.* Philadelphia: Franklin and Hall, iv, 32p.

Hayden, H. H. 1811. Geological Sketch of Baltimore, etc. *Baltimore Med. Philos. Lyceum* **1**:255-271.

Kain, John. 1818. Remarks on the Mineralogy and Geology of the Northwestern Part of the State of Virginia and Eastern Part of the State of Tennessee. *Am. Jour. Sci.* **1**:60-67.

Maclure, W. 1817. *Observations on the Geology of the United States of America.* . . . Philadelphia: J. Melish, x, 11-127, 2p., map, 2 plates.

Maclure, W. 1817. Observations on the Geology of the West India Islands, from Barbados to Santa Cruz, Inclusive. *Phila. Acad. Nat. Sci. Jour.* **1**:134-149.

Mease, J. 1807. *A Geological Account of the United States.* Philadelphia: Birch and Small, (8), 496, xivp., plates.

Mitchill, S. L. 1787. *Observations, Anatomical, Physiological, and Pathological on the Absorbent Tubes of Animal Bodies. To which Are Added Geological Remarks on the Maritime Parts of the State of New York.* New York: J. M'Lean, 16p.

Mitchill, S. L. 1814. Geology of Long Island. *Am. Mineralog. Jour.* **1**:261-263.

Mitchill, S. L. 1818. Observations on the Geology of North America; Illustrated by the Description of Various Organic Remains Found in that Part of the World. In *Essay on the Theory of the Earth* by G. Cuvier, pp. 319-431, 3 plates.

Pierce, J. 1820. Account of the Geology, Mineralogy, Scenery, etc. of the Secondary Region of New York and New Jersey, and the Adjacent Regions. *Am. Jour. Sci.* **2**:181-199.

Rafinesque, C. S. 1820. On the Geology of the Valley of the Mississippi. *Western Review and Miscellaneous Mag.* **2**:321-329.

Schoolcraft, H. R. 1820. On the Geology of Missouri, Arkansaw (sic.) and Illinois. *Plough Boy* **1**:45, 353.

Schöpf, J. D. 1787. *Beyträge zur mineralogischen Kenntniss des ostlichen Theils von Nord-Amerika und seiner Gebürge.* Erlangen, J. J. Palm, 195p.

Volney, C. F. C. B. 1804. *A View of the Soil and Climate of the United States, with Supplementary Remarks upon Florida; on the French Colonies on the Mississippi and Ohio, and in Canada; and on the Aboriginal Tribes of America.* Philadelphia: J. Conrad, xxviii, 446p., 3 plates, 2 maps.

Webster, J. W. 1820. Review of *An Index to the Geology of the Northern States* by Amos Eaton. *North Am. Rev.* **11**:225-239.

GEOLOGICAL REMARKS ON CERTAIN MARITIME PARTS OF THE STATE OF NEW YORK

In a Letter to Stephen Van Wyck, esq.

Samuel Latham Mitchill

Plandome, Queen's-county,
August 12, 1787.

Dear fir,

DURING my refidence in the country this fummer, it has been a principal part of my bufinefs to increafe my knowledge of phyfics, and to become a practical naturalift. I have on thefe accounts, walked over a confiderable tract of land, to examine with all poffible minutenefs, the phenomena which it afforded. My enquiries have been particularly directed to the difcovery of fomething ufeful, and, where this could not be attained, I have permitted myfelf to contemplate whatever of curious ftruck my notice. It has happened, in the courfe of fuch purfuit, that I have feen a fort of white clay which probably might be advantageoufly employed in the manufacture of porcelain—of a yellow argillaceous matter, that certainly would be ferviceable to the workers in leather—and a bright red ochre of iron, which is eafily mifcible with tar and oil, forms a good pigment for houfes, and doubtlefs could be applied to valuable purpofes by painters. I have found, befides, a quantity of martial pyrites, in feveral places, and of calciform iron ore in many others, fcattered along the fhores; but the bad quality of the former, and the fcanty quantity of the latter, render it unadvifeable to erect a furnace to extract the metal. Several chalybeate fprings gufh out, whofe water have been drank by valetudinarians, and may be profitable hereafter in medical cafes, where fuch practice is proper.

I have met alfo with fmall portions of the ferrum tubalcaini or bog ore of iron, on breaking which, fmall quartzy pebbles were found inclofed, proving it to be a fubftance of fecondary formation; and here likewife I may mention, that the petrifactions of wood and bark, which I have found, were always in an argillaceous ground, tinctured with iron, and fometimes mixed with flinty fand or gravel: but in thefe concretions, which are very frequent, I never have been able to find the leaft veftige of fhells, bones, or other animal relics, although thefe abound along the coaft, where fuch matters are plentiful.

During the time I made thefe remarks, and became poffeffed of fpecimens of each of the foffils enumerated, I was ftruck forcibly with a fet of appearances very different.

I obferved that the foffils found hereabout, in North Hamftead, are chiefly granitical, and the largeft rocks are compofed of quartz, fhoerl, and mica, in varied proportions, more or lefs intimately blended together, with now and then an admixture of feld fpath; in many places lie nodules of opaque quartz, either pure, or fometimes united to micaceous, and at others blended with calcareous matter; and pieces of fhapelefs quartz have occurred, on one part of which cryftals could be feen. Nodules of red jafper are frequently found, and I have feen it curioufly conjoined to quartz; chert, rag-ftone, and marble fometimes occur, but rarely; fhiftus may be found, but not plentifully; fhoerlaceous rocks are often met with, fometimes unmixed with any thing, but generally combined wtih quartz, mica, or garnet, and more rarely with filicious fand; the fhoerl is always cryftalized, and its colours are black, reddifh, and greenifh. I have feen hereabout a few pieces of freeftone, formed evidently by a cohefion of fea-fand; and on high grounds have difcovered fteatites and amianthus, and in low lands fibrous afbeftos in large collections. In a number of places, great bodies of fea-fhells may be found far above high-water mark; but thefe of Cowneck, as well as toofe of Matinicock, Newtown, and Rockaway, have evidently been carried up by the aboriginal inhabitants of the ifland, for in certain interftices one can difcover coals and afhes, and near them have been picked up, the ftone axes and arrowpoints formerly in ufe among the Indians, whofe bones may be eafily found by digging. No volcanic productions, fuch as lava, flag, or pumice-ftone, ever came under my obfervation.

I further obferved, that the arm of the fea, which feparates Long Ifland from the main land, although feveral miles in width, and deep enough to float large fhips, yet was fo interfperfed with fhoals, reefs, rocks, and iflands, that the navigation was crooked and difficult. Several of thefe are mere hills of rocky matter rifing above the furface of the water, and fome of them are wholly bare, while others are covered with fufficient foil to fupport a few trees, and fome fmaller fpecies of vegetables; their fubftance is of grey granite, intermingled with large fpots of white and reddifh quartz, fometimes pure, and fometimes mixed with fpar, mica, and feldfpath, and intersected with veins of different breadth, that often run in winding and serpentine courses, and are filled with the materials just named; the strata are vertical, or not much declining from the perpendicular, and their direction is from north-east to south-west, nearly.

Others of the islands are less solid, but have their shores covered with vast quantities of a like rocky matter, that is broken into smaller fragments; they have generally, as well as the former, bold shores, and their high banks of earth, undermined from time to time by the spring tides, are tumbling down—or, soaked by the rains, are wasting away.

On the adjacent part of the continent, the fossils are nearly of the same kinds, but the coast is in many places secured from further loss, by a firm lining of granite rock, disposed in perpendicular layers, or at most in an angle of eighty degrees to the horizon, and rising often in that manner, suddenly from the sea, or sinking as abruptly below it; the fissures are, in many cases, wide, and filled with the same sort of materials, as in the islands, and a similar course of the rocks from north-east to south-west is plainly to be seen, and even continues so for many miles to the northward and westward. In many places, I found separate masses of alumen plumosum, of shoerl variously coloured, and of black mica, scattered along the shore; but here I neither saw volcanic, metallic, or secondary fossils of any sort, except some of the concretions of quartzy gravel in ferruginous clay*. The coast of Long Island is generally sandy hereabout, but intersperfed with rocks and stones; that of the opposite continent exceeding solid and rocky, most of the moveable matter being washed away.

From the survey of the fossils in these parts of the American coast, one becomes convinced that the principal share of them is granitical†, composed of the same sorts of materials with the highest Alps†, Pyrenees, Caucasus, and Andes, and, like them, destitute of metals and petrifactions.

The occurrence of no horizontal strata, and the frequency of vertical layers, lead us further to suppose that these are not secondary collections of minerals, but are certainly in a state of primeval arrangement.

The steatites, amianthus, shoerl, feldspath, mica, garnet, jasper, shistus, asbestos, and quartz, must all be considered as primitive fossils, and by no means of an alluvial nature.

What inference remains now to be drawn from this statement of facts, but that the fashionable opinion of considering these maritime parts of our country as flats hove up from the deeps by the sea, or brought down from the heights by the rivers, stands unsupported by reason, and contradicted by experience?

A more probable opinion is, that Long Island, and the adjacent continent, were, in former days, continuous, or only separated by a small river, and that the strait, which now divides them, was formed by successive inroads of the sea from the eastward and westward, in the course of ages. This conjecture is supported by the facts which follow, to wit: 1. The fossil bodies on both shores have a near resemblance. 2. The rocks and islands lying between, are formed of similar materials. 3. In several places, particularly at White Stone and Hell Gate, the distance from land to land is very small. 4. Wherever the shore is not composed of solid rock, there the water continues to make great encroachments, and to cause the high banks to tumble down, as is true, not only here, but

NOTE.

* On the disposition of iron to form cements, see in the Swedish Memoirs for 1770, a paper by mr. Gadd, of which there is an abstract in Crell Chemisches Journal, 2. Th. Versuche mit dem Mortel, &c. 176, 8vo. Lemgo.—I have lately seen a curious instance of it around a piece of iron, found in the salt-water.

VOL. V.

NOTES.

† Dr. Shaw mentions the same sort of rocks as abounding in Arabia Petræa. See likewise Verbesserungen und Anmerckungen, &c. von derer Erden, und Steinen. By J. H. Pott. Potsdam, 1751, 4to. S. 47.

‡ Saussure, voyage dans les Alpes —and Kirwan's Geological Observations.

P

at Montock, Newtown, and elsewhere, at this very day. 5. The rocky piles in the Sound, called Executions, and Stepping Stones, and those named Hurtleberry Island, Pea Island, Heart Island, and many more, that lie up and down, are strong circumstances in favour of this opinion; for, from several of them, all the earthy matter, as far as the highest tides can reach, has long since been carried away; and, from the rest, the sand and gravel continue to be removed by daily attrition; as is true also of the Brothers, Ryker's, Blackwell's, and other islands. 6. There is a tradition among the race of men, who, previous to the Europeans, possessed this tract of country, that at some distant period in former times, their ancestors could step from rock to rock, and cross this arm of the sea on foot, at Hell Gate.

I have the honour to be,
With sincere esteem,
Yours, &c.
S. L. MITCHILL.

Observations on the Geology of the United States, explanatory of a Geological Map. By *William Maclure.*

Read January 20th, 1809.

NECESSITY dictates the adoption of some system, so far as respects the classification and arrangement of names the Wernerian appears to be the most suitable, First, Because it is the most perfect and extensive in its general outlines, and

[*Editor's Note:* The geological map accompanying this article is not included here. A colored reprint map from the 1817 edition is included in the reprinted (1969) edition of Merrill's *The First One-Hundred Years of American Geology.*]

secondly, The nature and relative situation of the minerals in the United States, whilst they are certainly the most extensive of any field yet examined, may perhaps be found to be the most correct elucidation of the general exactitude of that theory, as respects the relative position of the different series of rocks.

Without entering into any investigation of the origin or first formation of the various substances, the following nomenclature will be used.

Class 1st. *Primitive Rocks.*

1. Granite,
2. Gneiss,
3. Mica slate,
4. Clay slate,
5. Primitive Limestone,
6. Primitive Trap,
7. Serpentine,
8. Porphyry,
9. Sienite,
10. Topaz-Rock,
11. Quartz-Rock,
12. Primitive Flinty-Slate,
13. Primitive Gypsum,
14. White-Stone.

Class 2d. *Transition Rocks.*

1. Transition Limestone,
2. Transition Trap,
3. Grey Wacke.
4. Transition Flinty-Slate,
5. Transition Gypsum.

Class 3d. *Flœtz or Secondary Rocks.*

1. Old Red Sandstone or 1st Sandstone Formation,
2. First or Oldest Flœtz-Limestone,
3. First or Oldest Flœtz-Gypsum,
4. 2d or Variegated Sandstone,
5. 2d Flœtz-Gypsum,
6. 2d Flœtz-Limestone,
7. Third Flœtz-Sandstone,
8. Rock-Salt Formation,
9. Chalk Formation,
10. Flœtz-Trap Formation,
11. Independent Coal Formation,
12. Newest Flœtz-Trap Formation.

Class 4th. *Alluvial Rocks.*

1. Peat,
2. Sand and gravel,
3. Loam,
4. Bog iron ore,
5. Nagel fluh,
6. Calc-tuff,
7. Calc-sinter.

To the east of Hudson's river, the primitive class prevails, both in the mountains and the low lands, decreasing gradually as it proceeds south; it is bounded on the side of the ocean by the vast tracts of alluvial formation which skirt the great granite ridge, while it serves as a foundation to that immense superstructure of transition and secondary rocks, forming the great chain of mountains that occupy the interior of the continent to the westward.

The primitive to the eastward of Hudson's river constitutes the highest mountains, while the little transition and secondary that is found, occupies the low grounds. To the south of the Delaware, the primitive is the first rock, after the alluvial formation of the ocean, the lowest step of the stair, that mounts gradually through the different formations to the top of the Alleghanys.

To the eastward of the State of New-York, the stratification runs nearly north and south, and generally dips to the east, looking up to the White Hills, the most elevated ground; in New-York State, and to the southward and westward, the stratification runs nearly N. E. and S. W. and still dips generally to the east. All the rivers east of the Delaware, run nearly north and south, following the stratification, while the southern rivers incline to the S. E. and N. W. direction.

Throughout the greatest part of the eastern and northern States, the sea washes the foot of the primitive rock; commences the deposition of that extensive alluvial formation at Long-Island, increasing in breadth to the south, forming a great part of both the Carolinas and Georgia, and almost the whole of the two Floridas and Lower Louisiana. The coincidence of the Gulf-stream, with all its attendant eddies, depositions, &c. &c. rolling along this whole extent, from the Gulf of Mexico to Nantucket, may create speculative ideas on the origin of this vast alluvial formation, while the constant supply of caloric,* brought by that sweeping current from the tropics, may perhaps account for the sudden and great change in the temperature of the climate, within the reach of the Atlantic.

*About 100 miles S. E. of Nantucket, in the month of September, Fahrenheit's thermometer in the sea stood at 78°, while the air was only 66, and the sea in soundings 61.

The great distance occupied by the same, or similar substances, in the direction of the stratification, must strike the observer; as in the primitive rocks, the beds of primitive limestone and Dolomite (containing in some places crystallized felspar and tremolite) which are found alternating with Gneiss, for ten miles between Dover, State of New-York, and Kent, State of Connecticut, appear forty miles north, at Stockbridge, Connecticut, and eighty miles south, between Singsing and Kingsbridge, New-York; where, after crossing the Hudson river and dipping under the trap and sandstone formation in New-Jersey, they most probably re-appear in the marble quarries, distant from twelve to fourteen miles N. W. of Philadelphia,—a range of nearly 300 miles.

There is a bed of magnetic iron ore, from eight to twelve feet thick, wrought in Franconia, near the White Hills, New-Hampshire; a similar bed in the direction of the stratification, six miles N. E. of Phillipstown, on the Hudson river; and still following the direction of the stratification, the same ore occupies a bed of nearly the same thickness at Ringwood, Mount-Pleasant, and Suckasunny, in New-Jersey, losing itself as it approaches the end of the primitive ridge, near Blackwater,—a range of nearly 300 miles.

Instances of the same occur in the transition and secondary rocks; as the Blue ridge from the Hudson river to Dan river, consists of rocks of much the same nature, and included in the same formation.

That no volcanic productions have yet been found east of the Mississippi, is not the least of the many prominent features of distinction between the geology of this country and that of Europe, and may perhaps be the reason why the Wernerian system, so nearly accords with the general structure and stratification of this continent.

It is scarce necessary to observe, that the country must be considered of the nature of the *first* rock that is found in place, even should that rock be covered with thirty or forty feet of sand or gravel, on the banks of rivers, or in valleys; for example, the city of Philadelphia stands on primitive rock, though at the Centre-square, thirty or forty feet of sand and gravel must be penetrated, before the Gneiss rock, which ascertains the formation, is found.

Beginning at the bay of Penobscot (to the northward and eastward of which most probably the primitive descends through a gradual transition to the secondary, and thus into the Independent coal formation, found in such abundance in Nova Scotia;) and proceeding south, the sea coast is primitive to Boston, where the transition covers it as far as Rhode-Island.

ALLUVIAL FORMATION.

On the south east side of Long-Island the alluvial begins, occupying more than the half of that island; its western and northern boundaries are marked by a line passing near Amboy, Trenton, Philadelphia, Baltimore, Washington, Fredericksburg, Richmond, and Petersburg in Virginia, a little to the westward of Halifax, Smithfield, Aversborough and Parker's Ford on Pedee river, in North Carolina, west of Cambden near Columbia, Augusta on the Savannah river, Rocky Landing on the Oconee river, Fort Hawkins on the Oakmulgee river, Hawkinstown on Flint river, and running west, a little southerly, across the Chatahouchee, Alabama and Tombigby rivers, it joins the great alluvial bason of the Mississippi a little below the Natchez.

The ocean marks the eastern and southern limits of this extensive alluvial formation, above the level of which it rises considerably in the southern States, and falls to near the level of the sea, as it approaches the north.

Tide water in all the rivers from the Mississippi to the Roanoke stops at a distance from thirty to one hundred and twenty miles short of the western limits of the alluvial; from the Appomatox to the Delaware, the tide penetrates through the alluvial, and is only stopped by the primitive ridge.

The Hudson is the only river in the United States where the tide passes through the alluvial, primitive, transition, and into the secondary, in all the northern and eastern rivers, the tide runs a small distance into the primitive formation.

Through the whole of this alluvial formation, considerable deposits of shells are found; and a bank of shell limestone beginning in North Carolina, and running parallel to, and within the distance of from twenty to thirty miles of the edge of the primitive through South Carolina, Georgia, and part of the Mississippi Territory; in some places this bank is soft, with a large proportion of clay, in others, hard, with a sufficiency of the calcareous matter to be burned for lime, large fields of the same formation are found near cape Florida, and extending some distance along the coast of the bay of Mexico; in some situations, the calcareous matter of the shells has been washed away, and a deposit of silicious flint has taken their place, forming a porous flinty rock, which is used with advantage for mill-stones.

Considerable deposits of bog iron ore, occupying the lower situations, and many of the more elevated and dividing ridges between the rivers are crowned with a sandstone and puddingstone, the cement of which is bog iron ore.

Quantities of ochre, from bright yellow to dark brown are found in abundance in this formation, in flat horizontal beds alternating with other earths in some places; in others in kidney form masses from the size of an egg to that of a man's head, in form resembling much the flint found frequently in chalk formations.

PRIMITIVE FORMATION.

The south east limit of the great primitive formation is covered by the north western boundary of the alluvial formation from the Alabama river in the Mississippi Territory, (near which it is succeeded by the transition and secondary formations) to the east end of Long-Island, with two small exceptions; the first near Augusta on the Savannah river, and near Cambden in South Carolina, where a stratum of transition clay-slate intervenes, and from Trenton to Amboy, where the oldest sandstone formation covers the primitive along the edge of the alluvial.

From Rhode-Island (the greatest part of which is transition rock) to Boston, the primitive touches a transition formation, which most probably extends to the eastward, until it meets the alluvial along the sea coast by Elizabeth island, cape Cod

&c. &c. the eastern edge of the primitive from Boston to the bay of Penobscot is bounded by the ocean.

The north western boundary of this extensive range is marked by a line running to the eastward of lake Champlain, twenty or thirty miles westward of Connecticut river, to the westward of Stockbridge, twelve miles east of Poukepsy, skirting the high lands, then crossing the Hudson river at Philipstown, by Sparta about ten or fifteen miles east of Eastown, on the Delaware, three miles east of Reading on the Schuylkill, and a little west of Middletown on the Susquehannah, where it joins the blue ridge, and continues along it to Magotty Gap; from thence to four miles east of the lead mines at Austinville, and following a south western direction, by the stoney and iron mountains, six miles S. E. of the warm springs in Buncomb county, North Carolina, to the eastward of Hightown on the Cousee river, and a little to the westward of the Talapousee river, it meets the alluvial near the Alabama river, which runs into the bay of Mexico at Mobile.

In general the strata of this primitive rock run from a north and south to a north east and south west direction, and dip almost universally to the south east at an angle of more than 45 degrees from the horizon; the highest elevation is towards the north western limits, which gradually descends to the south east where it is covered by the alluvial, and the greatest mass, as well as the highest mountains, are found towards the northern and southern extremities of the north western boundaries.

The outline of the mountains of this formation is generally circular waving, in detached masses, with rounded flat tops, as the white hills to the north, or conically waving in small pyramidal tops, as the peaks of Otter, and the ranges of hills to the south; (has the climate any agency in the forms of the northern and southern mountains?) their height does not appear to exceed six thousand feet above the level of the sea, except perhaps the white hills, it is even probable that those mountains are not much higher.

Within the limits prescribed to this primitive formation, there is a range of secondary, extending with some intervals from the Connecticut to the Rappahannock rivers, in width generally from fifteen to twenty five miles, bounded on the north east from Connecticut river to New-Haven, by the sea, where it ends, to recommence on the south side of Hudson river; from Elizabeth town to Trenton, it touches the alluvial. From a little above Morrisville on the Delaware to Norristown, Maytown on the Susquehannah, passing three miles west of York, Hanover, and one mile west of Fredericktown, it is bounded by, or rather appears to cover a tongue of transition, which occupies progressively a diminishing width as far south as Dan river.

This secondary formation is interrupted after it passes Fredericktown, but begins again between Monocasy and Seneca creeks, the north eastern boundary crossing the Potomac, by the west of Centerville, touches the primitive near the Rappahannock, where it finishes. On the north west side it is bounded by the primitive, from some distance to the westward of Hartford, passing near Woodbury, and recommencing south of the Hudson, passing by Morristown, Germantown, &c. to the Delaware; after which it continues along the transition, by the east side of Reading, Grub's mines, Middletown, Fairfield, to near the Potomac, and recommencing at Noland's ferry, runs along the edge of the transition to the westward of Leesburg, Haymarket &c. to near the Rappahannock.

All this secondary appears to be the oldest red sandstone formation, though in some places about Leesburg, Reading &c. the red sandstone only serves as cement to a pudding, formed of limestone of transition, and other transition rock pebbles, with some quartz pebbles. Large beds of greenstone trap and wacke of different kinds, cover in many places this sandstone formation, and form the small hills, or long ridges which occur so frequently in it.

The stratification in most places runs from an east and west to a north east and south west course, and dips generally to the N. W. at an angle most frequently under 45 degrees from the horizon, covering both the primitive and transition formations, at every place where their junction could be examined; and in some places, such as the east side of the Hudson (where the action of the ater had worn away the sandstone) the smooth water-worn primitive was covered with large rolled masses of greenstone trap to a considerable distance, the hardness and solidity of which had most probably survived the destruction of their sandstone foundation; may not similar derangements be one of the causes of the broken and unconnected state of this formation?

Prehnite and zeolite are found in the trap of this formation; considerable *deposits* of *magnetic iron ore* at Grubb's mines are enveloped, and have their circular layers intersected by greenstone trap, on a ridge of which this extensive cluster of iron ore appears to be placed.

Grey copper ore has been found in the red sandstone formation near Hartford and Washington in Connecticut; at Scheuyler's mines in Jersey, *copper pyrites* and *native copper* have been found. The metallic veins on Perkiomen creek, containing *copper pyrites*, *blende*, and *galena*, are in the same formation; running nearly north and south, across the east and west direction of the red sandstone; a small bed from an half to three inches thick of *brown* or *tile copper ore* is interspersed and follows the circular form of the iron beds at Grubb's mines.

Besides the sandstone formation, there is included within the described limits of the primitive, a bed of transition rocks, running nearly S. W. from the Delaware, to the Yadkin river, dipping generally to the south east 45 degrees or more from the horizon; its width is from two to fifteen miles, and runs from the west of Morrisville, to the east of Norristown, passes Lancaster, York, Hanover, Fredericktown, Bull run mountain, Milton, foot of Pig river, Martinsville, and finishes near Mount Pilot, between the Delaware and Rappahannock; it is partially covered by the red sandstone formation, and is in the shape of a long wedge, the thick end, touching the Delaware, and the sharp end, terminating at the Yadkin.

This range consists of beds of blue, grey, red and white small grained transition limestone, alternating with beds of *grey wacke* and *grey wacke-slate;* with granular quartzose rocks, and a great variety of transition rocks, not described or named in any treatise yet published; much of this limestone is intimately mixed with *grey wacke-slate*, other portions of it contain so great a quantity of small grained sand, as to resemble *Dolomite*, and perhaps might with propriety be called the *transition Dolomite*, in many places veins and irregular masses of silex, variously coloured (mostly black) run through it, and considerable beds of fine grained white marble, fit for the statuary, occur.

Limestone spar runs in veins and detached masses, through the whole of this formation, both it, and the *grey wacke-slate* contain quantities of *cubic pyrites;* galena has likewise been found near Lancaster, and many veins of the *sulphate* of *barytes* traverse this formation, which runs about 25 to 30 miles south east, and nearly parallel to the great transition formation. A similar formation, about fifteen miles long, and two to three miles wide, occurs on the north fork of the Catabaw river, running along Linville and John's mountains, near to the Blue ridge; a bed of transition rock, commencing on Green pond mountain, Jersey, runs through Suckasunny plains, increasing in width as the primitive range decreases, until it joins the great transition formation between Easton and Reading.— On the west side of this partial transition formation, from the Potomac to the Cataba, between it and the great western transition range, a series of primitive rocks intervenes, something different from the common primitive, having the structure of *gneiss;* with little *mica,* the scales of which are detached and not contiguous; much *felspar,* rather granular than crystallized; *mica-slate,* with small quantities of *scaly mica;* clay-slate, rather soft and without lustre, the whole having a dull earthy fracture, and gritty texture, partaking of transition and primitive, but not properly belonging to either; this rock is always found on the edge of the primitive, before you come upon the transition, but no where in such quantities as in this range; there are many varieties of it, so that it imitates almost every species of the common primitive rocks, but differing from them, by having a dull earthy fracture, gritty texture, and little or no crystallization.

About ten or twelve miles west of Richmond, in Virginia, there is an *independent coal formation,* twenty to twenty five miles long, and about ten miles wide, it appears to be not far distant from the range of the red sandstone formation, it is situated in an oblong bason accompanied by whitish *freestone, slaty clay, &c.*

with *vegetable impressions*, as well as most of the other attendants of that formation; this bason lays upon, and is surrounded by primitive rocks. It is more than probable that within the limits of so large a mass of primitive, more partial formations of secondary rocks may be found.

A great variety of mineral substances is found in this primitive formation, such as *garnets* in the *granite*, from the size of a pin head to the head of a child; *staurotide; andalusite; epidote* in great abundance; *tremolite;* all the varieties of *magnesian rocks; emerald,* touching *graphic granite* and disseminated in the *granite* of a large extent of country; *adularia; tourmaline; hornblende; sulphate* of *barytes; arragonite* &c.

From the number already found, in proportion to the little research that has as yet been employed, there is every reason to suppose, that in so great an extent of crystalline formation, almost every mineral which has been discovered in similar situations on the ancient continents, will be found on this.

The metallic substances which are found in this primitive, are generally extensive like the formation. *Iron pyrites* run through vast fields, principally of *gneiss,* and *mica-slate; magnetic iron ore* forms vast beds, from ten to twelve feet thick, generally in a *hornblende* rock, occupying the higher elevations, as at Franconia, high lands of New-York; the Jerseys; Yellow and Iron mountain, in the west of North Carolina, &c. &c. Black, brown, and red *hematitic iron ores* are found in Connecticut and New-York, &c. *Crystals* of *octahedral iron ore* are disseminated in *granite* (some of which have polarity, as at Brunswick) and in many varieties of the *magnesian genus; black lead* exists in beds from six to twelve feet wide, traversing the States of New-York, Jersey, Virginia, Carolina, &c. *Native* and *grey copper ore* occur near Stanardsville and Nicholson's Gap, disseminated in a *hornblende* and *epidote rock,* bordering on the transition; *molybdena* is found at Brunswick, Maine; Chester, Pennsylvania; Virginia; North Carolina, &c. *Arsenical pyrites* have been discovered in large quantities in the district of Maine; *rutile,* and *menachanite* exist in a large bed, on the edge of the primitive near Sparta, in Jersey, having a large grained marble, with *menachanite* and *negrine* imbeded in it on one side, and *hornblende* rock on the other; this bed contains likewise large quantities of *blende*; *detached* pieces of *gold* have been found in the beds of some small streams in North Carolina and other places, apparently in a *quartz* rock. *Manganese* has been found in New-York, North Carolina, &c. Near the confines of the red sandstone and primitive formations, a *white ore* of *Cobalt* has been worked above Middletown on the Connecticut river, and it is said near Morristown in New-Jersey.

The general nature of metallic repositories in this formation appears to be in beds, disseminated or lying in masses; when in beds (as the *magnetic iron ore*, and *black lead*) or disseminated as the *iron pyrites, octahedral iron ore, Molybdena*, &c. they occur at intervals through the whole range of the formation; veins to any great extent have not yet been found in this formation.

TRANSITION FORMATION.

This extensive field of transition rocks, is limited on the S. E. side from a little to the eastward of lake Champlain, to near the river Alabama, by the N. W. boundary prescribed to the primitive rocks; on the N. W. side it touches the S. E. edge of the great secondary formations, in a line, that passes considerably to the westward of the dividing ridge, in Georgia, North Carolina, and part of Virginia, and runs near it in the northern parts of that State, and to the eastward of it in the States of Pennsylvania and New-York.

This line of demarkation runs between the Alabama and Tombigby rivers, to the westward of the north fork of the Holstein river, until it joins the Alleghany mountains near the sulphur springs, along that dividing ridge to Bedford county in Pennsylvania, and from thence N. E. to the east side of the Catskill mountains on Hudson's river. This line of separation of the transition and secondary formations, is not so regularly and distinctly traced as in the other formations, many large valleys are formed of horizontal secondary limestone, full of shells, while the ridges on each side consist of transition rocks, &c. the two formations interlock, and are mixed in many places, so as to require much time and attention to reduce them to their regular and proper limits. It is however probable, that to the N. W. of the line here described, little or no transition will be found, although to the S. E. of it, partial formations of secondary may occur.

The breadth of this transition formation is generally from 20 to 40 miles, and the stratification runs from a north and south to a north east and south west direction, dipping generally to the N. W. at an angle in most places, under 45 degrees from the horizon. On the edge of the primitive; it, in some places, deviates from this general rule, and dips for a short distance to the south east. The most elevated ground is on the confines of North Carolina, and Georgia, along the S. E. limits to Maggotty Gap, descending towards the N. W. until it meets the secondary; from Maggotty Gap, north easterly, the highest ground is on the north west side, sloping gradually towards the primitive, which ranges along its south east boundary.

The outline of the mountains of this formation, is almost a straight line, with few interruptions, bounding long parallel ridges of nearly the same height, declining gently towards the side, where the stratification dips from the horizon, and more precipitous on the opposite side, where the edge of the strata comes to the surface.

This formation is composed of a small-grained transition limestone, of all the shades of colour from white to dark blue, and in some places it is red, intimately mixed with grains of *grey wacke slate*, also of *lime spar* in veins, and disseminated; *siliceous flinty* veins and irregular masses, in many places there is an intimate mixture of small sand, so as to put on the appearance of *dolomite*, this is in beds from 50 to 5000 feet in width; it alternates with *grey wacke*, and *grey wacke slate*, a siliceous aggregate, having particles of a light blue colour, from the size of a pin head to that of an egg, disseminated, in some places in a cement of a slaty texture, and in others in a quartz cement; a fine sandstone cemented with quartz in large masses, often of a slaty structure, with small detached scales of *mica* intervening, and a great variety of other rocks, not described or named by any author, which from their composition and situation cannot be classed but with the transition.

K k

The limestone, *grey wacke,* and *grey wacke slate,* generally occupy the vallies; the quartzose aggregates, the ridges; amongst which is that called the *millstone grit;* this must not be confounded with another rock, likewise denominated the *millstone grit,* which is a *small grained granite,* with much quartz, found in the primitive formation; there are many and extensive caves in the limestone of this formation, where the bones of many animals are found, as well as the remains of marine insects and shells.

Beds of *coal-blende,* accompanied by *alum slate* and *black chalk,* have been discovered in this formation on Rhode Island; the Leheigh and Susquehannah rivers; (a large body of *alum slate* which occurs on Jackson's river in Virginia is perhaps only a part of a similar formation;) *powerful* veins of the *sulphate* of *barytes* cross it, in many places it is granular, as that near Fincastle; or slaty, as in Buncomb county, North Carolina.

Iron and *lead* have as yet been the principal metals found in this formation; the *lead* in the form of *galena,* in clusters, or what the Germans call *stock-werck,* as at the lead mines on New river, Wythe county, Virginia; the *iron* is disseminated in the form of pyrites; hematitic and *magnetic iron ores,* and considerable quantities of the *sparry iron ore* occur in beds and they are likewise disseminated in the limestone.

SECONDARY FORMATION.

The south east limit of this extensive formation is bounded by the irregular border of the transition, from between the Alabama and Tombigby rivers, to the Catskill mountains. On the north west side it follows the shore of the great lakes, and loses itself in the alluvial of the great bason of the Mississippi, occupying a surface from 200 to 500 miles in breadth.

Its greatest elevation is on the south east boundary, from which it falls down, almost imperceptibly, to the north west and mingles with the alluvial of the Mississippi, having an outline of mountain, straight and regular, bounding long and parallel ranges of a gradually diminishing height as they approach the N. W. limits. An almost horizontal stratification, or the strata waving with the inequalities of the surface, distinguishes this from the two preceding formations.

Immense beds of secondary limestone, of all the shades from light blue to black, intercepted in some places by extensive tracts of sandstone and other secondary aggregates, appear to constitute the foundation of this formation, on which reposes that great and valuable formation, called by Werner the *independent coal formation*, extending from the head waters of the Ohio, with some interruptions, all the way to the waters of the Tombigby, accompanied by its several usual attendants, *slaty clay* and *freestone* with vegetable impressions &c. but *in no instance* that I have seen or heard of, is it *covered* or does it *alternate* with any rock resembling *basalt*, or indeed any of those called the newest *flætz trap formation*.

Along the S. E. boundary, not far from the transition, a *rock-salt* and *gypsum* formation has been found; on the north fork of Holstein not far from Abington, and on the same line south west from that in Green county and Pidgeon river, State of Tennessee, it is said considerable quantities of *gypsum* have been discovered; from which, and the numerous salt licks and salt springs which are found in the same range, as far north as lake Oneida, it is probable, that this formation is on the same great scale, which is common to all the other formations on this continent: at least rational analogy supports the supposition, and we may hope one day to find, in abundance, those two most useful substances, which are generally found mixed or near each other in all countries that have been carefully examined.

The metallic substances which have been already found in this formation, are *iron pyrites*, disseminated, both in the coal and limestone; *iron ores*, consisting principally of *brown, sparry* and *clay iron stone*, in beds; *galena*, whether in veins or beds is not ascertained. The large deposits of *galena* at St. Louis on the Mississippi, have been described as detached pieces, found covered by the alluvial of the river, of course not in place; all the large specimens which I have seen, were rolled masses, this rather confirms the opinion, that they were not found in their original places.

On the great Kanawa river, near the mouth of Elk river, there is a large mass of black (I suppose vegetable) earth, so soft, as to be penetrated by a pole from 10 to 15 feet deep; out of the hole thus made, a stream of *hydrogene gas* frequently issues, which will burn for some time. In the vicinity of this place there are *constant streams* of that *gas*, which it is said when once lighted will burn for weeks, a careful examination of this place, would probably throw some light on the formation of coal and other combustible substances, found in great abundance in this formation.

From near Kingston on lake Ontario, to some distance below Quebec (as far as I can recollect, not having my note-book here) it is principally primitive; and from all the information I could collect, that great mass of continent, lying to the north of the 46th degree of latitude, for a considerable distance to the west, consists mostly of the same formation; from which it is probable, that on this continent, as well as in Europe and Asia, the northern regions are principally occupied by the primitive formation.

The foregoing observations are the result of many former excursions in the United States, and a knowledge lately acquired by crossing the dividing line of the principal formations, in 15 or 20 different places, from the Hudson to Flint river; as well as from the information of intelligent men, whose situation and experience, make the nature of the place near which they live familiar to them; nor has the information that could be acquired from specimens, when the locality was accurately marked, been neglected, nor the remarks of judicious travellers.

Notwithstanding the various sources of information, much of the accuracy of the outlines of separation between the formations, must depend on rational analogy; for instance, between Maggotty and Rockfish Gaps, a distance of upwards of sixty miles, I found in six different places which were examined, that the summit of the blue ridge divided the primitive and transition formations: I of course concluded, that in places where I had not examined (or which from their nature could not be examined) the blue ridge from Maggotty Gap to Rockfish Gap, was the boundary of the two formations.

The map of the United States on which those divisions are delineated, though I believe the best yet published, is exceedingly defective in the situation and range of mountains, courses and windings of rivers &c. but as the specimens which I collected every half mile, as well as the boundaries of the different formations, are from the positive situations of the different places, the relative arrangement of the map cannot change them, but must become more exact, as the geographical part is made more accurate.

In adopting the nomenclature of Werner, I do not mean to enter into the origin or first creation of the different substances, or into the nature and properties of the agents which may have subsequently modified or changed the appearance and form of those substances; I am equally ignorant of the relative periods of time in which those modifications or changes may have taken place; such speculations are beyond my range, and pass the limits of my inquiries. All that I mean by a formation, is a mass of substances (whether adhesive, as rocks; or separate, as sand and gravel;) uniform and similar in their structure and relative position, occupying extensive ranges, with few or no interpolations of the rocks belonging to another series, class, or formation; and even where such partial mixtures apparently take place, a careful examination will seldom fail to explain the phenomenon without shaking the general principle, or making it a serious exception to the rule.

In the account of the metals and minerals, it is not intended to give a list of the number, extent and riches of the metallic and mineral repositories; the nature of the ore or mineral, with a description of its relative position, in regard to the surrounding substances, is the principal object of geology, which cannot be understood by microscopic investigation, or the minute analysis of isolated rocks and detached masses; this would be like the portrait painter dwelling on the accidental pimple of a fine face: the geologist must endeavour to seize the great and prominent outlines of nature; he should acquaint himself with her general laws, rather than study her accidental deviations, or magnify the number and extent of the supposed exceptions.

which most frequently cease to be so when judiciously examined.

Should this hasty and imperfect sketch, call forth the attention of those possessed of more talents and industry for the accurate investigation of this interesting subject, the views of the writer will be fully accomplished.

AN AMENDMENT PROPOSED TO THE GEOLOGICAL CHART OF THE UNITED STATES, AS RESPECTS THE CHARACTER OF THE NORTH SIDE OF LONG-ISLAND, WHICH IS SHOWN TO BE ALLUVIAL AND NOT PRIMITIVE, AS THEREIN STATED

Samuel L. Mitchill

NEW-YORK, July 4th, 1811.

DEAR SIR,

IN the sixth volume of the Philadelphia Philosophical Transactions, is an able and instructive memoir by William M'Clure, Esq. on the geology of the United States; accompanied by a map, coloured in such a manner as to present the different formations strongly to the eye. They are the same which you showed me lately, as having been sent you in the form of a pamphlet, by one of your correspondents in Paris.

In these I observe, that the north side of Long-Island is represented as *primitive*, while the south is described as *alluvial*. The ridge of hills which extends the greater part of its length, from the west to the east, is the barrier.

Of the correctness of the memoir and chart, as far as the alluvial character of Mattawacks is concerned, I entertain no doubt. But I have reason to suppose, that the portion of that island, which has been delineated as primitive, is, with very moderate exception, alluvial also.

Formerly, indeed, I was of a different belief. In my report to the Agricultural Society, in 1796, on the mineralogical history of New-York, I classed this among the primeval islands. I was induced so to do, on account of the masses of loose and detached granite, schœrl, and asbestoid, strewed over the region situated between the hilly ridge and the Sound; the dissimilarity of that soil, and configuration from the tract lying between the ridge and the ocean; the greater resemblance which the north side of the island bears to the contiguous continent; and above all, the strata of granite and gneiss which occur at and near Hurlgate, on the Long-Island-side of that remarkable pass.

Further observation has led me to suppose that both sides of Long-Island are alluvial, with the exception of the granitical layers before mentioned, being in Newtown: and I deem it but just to myself and the subject, as well as to those who may have been misled by following my former opinion, to state the reasons which have induced me, in some degree, to change it.

The first piece of evidence was derived from an examination of the strata penetrated by a well-digger, in the town of North-Hempstead, in 1804. The place where this opening was made, is in the land of Mr. G. Rapalje, on the road extending across the head of Cow-Neck, between Great-Neck and Hempstead-Harbour. The well was eighty feet deep; and the appearances were as follow:

1st. Sandy loam, or a mixture of clay and small gravel—10 feet.

2d. Compact clay—1 foot.

3d. Sand and gravel—9 feet.

4th. Gravel and roundish stones, chiefly of quartz and granite, breccias and argillaceous iron-ore, of the size and form of paving-stones—10 feet.

5th. Masses of grey granite jumbled together—5 feet.

6th. Gravel and roundish stones, as in the 4th stratum—3 feet.

7th. Hard loam—3 feet.

8th. Soft clay—six inches.

9th. Sea-sand, resembling that found on the shores of tide-water; sharp grit; coarse gravel, and rounded or water-worn stones, variously mixed, and alternately deposited—38 1-2 feet.

10th. Water—80 feet deep.

The second proof was afforded by an inquiry into the stratification towards the west end of the island, in the town of Brooklyn. The result of an examination of the materials thrown out of Mr. Johnson's well, eighty-four feet deep, in a spot between Wallaboght and Guanas, during April, 1811, was this:

1st. Sandy loam—3 feet.

2d. Hard concretion, requiring the pick-axe to break it up; formed chiefly of clay, sand, and stones, and strongly tinctured with iron of a yellowish red colour; and intermingled with rocks of gneiss, hornblende, and brittle ferruginous slate—15 feet.

3d. Loose gravel, and greyish sand mixed with thin streaks of gravel; the sand coarse, quartzy, and of a sharp grit; the gravel mixed with small masses of white quartz, and black touch-stone, with now and then breccias, micaceous slate and ferruginous sand-stone—20 feet.

4th. Alternate layers of sand and gravel, not more than two or three feet thick, containing sometimes rounded pieces of coarse green soap-stone, in addition to the materials in the stratum immediately above it—17 feet.

5th. After having thus descended fifty-five feet, strata of sand and gravel occur by turns. They do not respectively exceed three or four feet in thickness, and contain marine shells, mostly of clams and oysters. And it is remarkable that these animal remains are never found in the strata of sand, but uniformly in those of the gravel only—29 feet.

6th. Water, at the depth of 84 feet.

The third consideration in favour of the alluvial constitution of the north side of Long-Island, is deduced from the frequent occurrence of animal and vegetable substances at considerable distances below the earth's surface. I shall state to you a few, out of a great number of cases.

The shells of *clams* and *oysters* are found almost universally, sixty feet under ground, in Brooklyn, New-Utrecht, Flatbush and Newtown. A *periwinkle shell* was discovered forty-three feet down, at New-Utrecht. In Newtown, *carbonated wood*, sometimes by itself, and sometimes incrusted by pyrites, was raised from the bottom of a shaft fifty feet deep. In Bushwick, at the depth of forty-five feet, they found *the body of a tree* lying across the well they were digging, and they cut through it, rather than abandon the job. I say nothing of the *wood* discovered sixty feet deep, a little to the eastward of Westbury meeting-house: nor of *the bark, and other parts of a tree*, raised from the depth of forty feet at Eastwoods; because both these places are situated to the south of the barrier-ridge, and are within the district allowed by all to be alluvial. But I will mention the fact, of *wood* found at Success, on the very summit of the hills, thirty feet beneath the sward.

These facts appear to be conclusive on the subject. Yet there are two difficulties to be overcome. The first is, the *stratified granite* on the shore of Long-Island, at Hurlgate and its neighbourhood; and the second, the *detached masses of the like material*, varying from the bulk of two or three feet square, to the magnitude of the *mill-stone-rock*, a solitary block which I have computed elsewhere, (3 Med. Rep. p. 330) to contain twenty thousand and four hundred cubic feet.

The stratified granite of Long-Island, reaches not many miles along the shore, and that only between Flushing Bay and Newtown Creek. It seems to extend from the shore toward the interior, no great distance; apparently but a few rods.

In contemplating the loose and nodular rocks of primeval date, it is not easy to assign a perfectly satisfactory reason for their appearance, in such a state of dispersion, among materials of recent formation. But it is a fact that alluvial strata, in the very city and vicinity of New-York, are superinduced upon the primitive. Stratified granite, gneiss and steatites, constitute the old basis of the Manhattan island; and yet in digging away Bayard's hill, the heights near Corlear's Hook, and the other natural elevations, which are yet partially in being, we uniformly observe horizontal strata abounding in water-worn stones, which are very serviceable in paving the streets. These alluvial deposits overlay the more ancient beds of rocky matter. They may fairly be considered as the ruins of older strata, or as derived therefrom. And if, in the changes that the materials have undergone, some fragments have not been worn away nor reduced so small as others, but still retain considerable bulk, their existence can, nevertheless, be reconciled to geological principles. The same cause which was powerful enough to arrange the alluvial layers, as we find them, may have brought along detached and solitary parcels from their original abodes.

On the whole, the only primitive strata upon Long-Island exist at Hurlgate; and there they form but a very narrow strip; a mere margin of the shore, for the distance of between four and five miles. This being the fact, I hope our ingenious and enterprising friend may soon receive information about it, and amend the next edition of his chart accordingly.

While I congratulate you on the rapid progress of mineralogical and geological researches in our country, permit me to renew the assurance of my particular esteem and consideration.

SAM. L. MITCHILL.

REVIEW OF Observations on the Geology of the United States of America; With Some Remarks on the Effects Produced on the Nature and Fertility of Soils, by the Decomposition of the Different Classes of Rocks, and an Application to the Fertility of Every State in the Union, in Reference to the Accompanying Geological Map—With Two Plates, by William Maclure

C. S. Rafinesque

SEVERAL years ago Mr. Maclure communicated to the Philosophical Society of Philadelphia, some observations on the geology of the United States; he has now somewhat enlarged and corrected his former memoir, increasing it at the same time with an attempt to apply geology to agriculture, in which he is highly commendable, as we have no doubt that his endeavours will be found practically useful, even by those who do not entertain any high idea of scientific researches. Every science is connected with the wants of mankind; and many sciences are indebted for their origin to those wants, which increase in proportion to civilization and refinement. Agriculture sprung from the inadequacy of nature's spontaneous supplies of food for a large population, and has but lately become a science; medicine sprung from the natural desire of relieving our pains and lengthening our lives; geometry from the necessity of ascertaining the extent and limits of our fields; geography from the importance of knowing the strength and resources of our own country, and the means and dispositions of our neighbours; astronomy from the exigences of shepherds and navigators; physics from the need of becoming acquainted with the phenomena which surround us, as well to avail ourselves of their co-operation, as to avert some of the dreadful disasters of which they are sometimes the cause; cosmony from the cravings of nature, which instigate us to learn what animals, plants, or minerals may be made subservient to our use, or afford us food, raiment, weapons, tools, &c.

All the divisions of knowledge to which we have given the names of arts or sciences, have, therefore, a common origin—our wants! a common object—our uses! a common view—our improvement! These selfish motives are those which govern the majority of mankind; but philosophy refines and elevates them. This common origin and object of the sciences has often led to the belief of their identity as if they were all concentrated in a universal science. This hypothesis cannot now have many adherents, since the different scientific pursuits have been so well illustrated and distinguished; yet every one must be aware of the intimate connexion which exists between all the sciences. For instance, botany and geometry, which appear so widely distinct, are yet so far connected that botany must borrow part of its language from geometry, and geometry some of its forms from botany.

In a peculiarly improved stage and extended state of the sciences, the necessity of dividing them into minor sciences or branches begins to be felt, and such a division usually takes place shortly afterwards. It is to such a period that we are indebted for the new science of geology, or the knowledge of the solid part of the earth. This science was for a long time blended with natural history, mineralogy, astronomy, cosmogony, mythology, history, to which it is more or less connected, without properly belonging to either; but it has in recent days been raised to the dignified station of a separate science, and can already number among its votaries such men as Cuvier, Werner, Hutton, Patrin, Lametherie, &c. in Europe, while in the United States many enlightened men do not disdain to cultivate it for the benefit of the present generation and of posterity.

Among the latter Mr. Maclure stands conspicuous for zeal, assiduity, perspicuity, liberality, utility, and an early attention to this important subject. It is not by the size of his work that we must judge of its value; but by its intrinsic merit. We believe that in the small number of pages of his volume, more essential facts and useful truths are disclosed than in many thick volumes of yore. We shall endeavour to collect such of them as our limits will allow, and such that a tolerable idea of the value of his observations may be formed; and the few imperfections which we may have occasion to notice, will but slightly invalidate its real merit.

We agree altogether with our worthy author, when he states the fallacy of the numberless presumptive theories of the earth, which have so often been set up. While we have scarcely studied one-fourth part of the *surface of the earth*, and while the interior of our globe is totally unknown, all speculative theories must be considered as the *novels of geology* rather than its history. How many of them have even been founded upon a few local facts, which are belied by so many different facts elsewhere! Mr. Maclure mentions that those animals whose bones have been found in northern climates, while they (or their congenerous species) are now found only in tropical climates, might have been migratory, as the wild Buffaloe of America is at this time;—he might have added, that most of them being different from the now living species, were probably (as the mammoth of Siberia was to a certainty) covered with a thick fur suitable to the climates they dwelt in. Yet to account for this simple fact, a supposition has been advanced, that the equator was once where the poles are now, and vice versa! If the mutation of the poles could only be supported by this false reasoning, every supposition of the kind would fall to the ground. Fire and water were, till lately, considered as the only agents acting over the earth,—now galvanism is allowed to have also its share; but electricity, magnetism, light, gases, air, frost, compression, and animal and vegetable agency, &c. have certainly also their share; wherefore every theory founded upon a simple or single agent, becomes an erroneous system.

Our author adopts Werner's classification of rocks; but he is not satisfied with his distinctive names of primitive and secondary; he might have added his transition, which denomination is certainly illusive. The fact is, that there are but four formations of rocks and earths, *all of which*, even granite, are stratified; they are the crystallized, the deposited, the volcanic, and the organic formations; the first originates in crystallizations, the second in depositions, the third in emissions, and the last in organic remains; if a fifth formation was to be added, it ought to be the agglomerated formation. The transition formation belongs to all the formations in various instances, and the alluvial to the deposited formations. All these formations often happen to be blended, which destroys altogether the theories of universal separate formations, since suppositions must yield to facts; and strata vary from the thickness of a sheet of paper to the immense thickness of several thousand feet, so far as they have been penetrated or seen.

The uniformity of the formations in the United states, and the regularity of their dispositions, strike every observer who has witnessed the disparity and irregularity which are exhibited in the formations of Europe. Mr. Maclure traces an able parallel between the two continents, and describes next the outlines and limits of the formations, rocks, mountains and strata of our continent, being the result of nearly thirty different excursions across their nucleus, which runs from northeast to southwest. He describes the whole in general results, disdaining minute investigation of insulated rocks and detached masses: yet if there are some of such, which may throw light upon the approximating formations, why should we neglect them altogether? We shall not follow him through his leading remarks, and his divisions; a single glance at his map will convey a better idea of his principles, the results of which are, that nearly all the New-England states, the northern part of New-York, and a broad stripe as far as Georgia, are primitive; that the alluvial formation extends from Long-Island to Louisiana, from the Atlantic to the granite up the Mississippi as far as the mouth of the Ohio; that the limestone, or secondary formation, extends all over the western states, as far as the lakes, including most of New-York, and that it is divided from the primitive by a transition region. A formation of sandstone exists in the primitive, in New-York, Maryland, Connecticut, &c.

Notwithstanding the able researches of our author, we cannot but regard his results, as well as those of Volney, as mere attempts towards the knowledge which he means to convey; we know of several instances in which the limits assigned to some formations are not altogether correct, nor can they ever be completely known, but after a series of long, minute local observations all over the United States; and even then, how are we to know when those limits are absolute or relative? We would advise observers to notice the angle of inclination of the strata at the place of their disappearance, whence a probable calculation may be made of their further depth and extent. A long period must elapse before we can acquire a complete knowledge of the soil we inhabit; we must sink wells and shafts, dig mines and coal-pits to great depths, ere we can assert which is the predominant formation in the strata we tread upon; but we must especially collect and describe all the organic remains of our soil, if we ever want to speculate with the smallest degree of probability, on the formation, respective age, and history of our strata. Mr. Maclure has altogether omitted these accessories or auxiliaries, which have received, with much propriety, the name of *medals of nature:* he says little or nothing of the numberless animal remains, shells, polyps, &c. found all over our deposited and agglomerated soils, or alluvial, limestone, sandstone regions. He omits the alluvial found in Ohio and New-England, &c. The regions north of the lakes are a blank in his map; they are probably of primitive or granitic formation. The present great lakes of North-America, and those which have to a certainty existed elsewhere in ancient times, have had more influence on some parts of the soil than he is aware of. He has not mentioned any volcanic soils and rocks in the United States; yet there are certainly some, which he has classed, with the Wernerian school, among transition and secondary; but the trap, wake, coal, and clay formations, which are found in many parts, are here, as in Europe, evidently of volcanic, or emitted formation. Volcanoes do not always emit fire and lava, nor heap up mountains and craters; they often vomit water and mud, and, when they are covered by water, their smoke and ashes form, under the water, strata of various substances: such have been the ancient submarine volcanoes of Connecticut, New-York, Pennsylvania, Virginia, Alabama, &c.

The second plate of this work contains five transverse sections of the United States: 1. across lake Champlain and the White Hills; 2. from Plymouth to lake Erie; 3. from Egg-Harbour to Pittsburg; 4. from Cape Henry to Abingdon; 5. from Cape Fear to the Warm Springs. They give a tolerably good idea of the succession of formations; but we hope, that by leading each formation to the level of the sea, it was not meant to imply that they really reach it, else we should ask how was it known to be so?

We now proceed to the second part of this work, or the practical part thereof, wherein the author relates, with much propriety in the preface, how various are the practical results to be derived from the study of geology; it is by such a study that we are safely guided in our search for coal, salt, gypsum, limestone, sandstone, millstones, grindstones, whetstones, marble, clay, marl, slate, ores, &c. For instance, those who should search for coal in a primitive region, or under granite, would lose their time and money: those who mistake pyrites and mica for ores, find soon their delusion to their cost. It will teach you to pave turnpikes with quartz, which will wear two years, instead of limestone or any soft stone, which will not last three months. When clay contains too much calcarious matter, it cannot make good bricks, and when limestone contains too much argillaceous matter, it cannot make lime.

The theory of the decomposition of rocks is treated with great ability and perspicuity; it is worth while for every enlightened agriculturalist to become acquainted with it: the results are, that the best soils for agricultural purposes are those proceeding from the decomposition of wake, limestone, lava, tuffa, &c. that the worst are those resulting from clay, salt, sand, quartz, &c. that alluvial and transition formations partake of such formations as they have been washed from; that vegetable mould is the common manure of nature, that gypsum is the next, marl and clay, of sand, and vice versa, &c.

In the last chapter Mr. Maclure enters at length into an investigation of the probable effects which the decomposition of rocks may have on the nature and fertility of the soils of the different states of North-America, when such soils are in their pristine state, since, when covered with vegetable and animal manure or mould, their fertility lasts as long as such mould remains. In result it appears that Pennsylvania and New-York possess the greatest quantity of good lands among the Atlantic states, while all the western states enjoy an equal fertility, being all situated in the limestone formation. All the alluvial region fronting the ocean appears to possess a peculiar character, the soil being almost every where light, dry and sandy, or swampy; this soil, when mixed with marl, which is generally found under it, forms a good cultivable ground. It is probable that cotton, the staple produce of this region south of the Chesapeake, will, at a future period, be found suitable to the whole region, and cultivable as far north as Long-Island, and on those Hempstead plains, now thought almost unfit for cultivation, as were formerly thought the pine barrens of South-Carolina.

Mr. Maclure indulges sometimes in digressions in which some happy thoughts are discernible: his great division of the states, into states east and west of the Alleghany, is quite natural, and the probable consequences of their respective features are truly delineated. Happily the Atlantic states are divided also naturally in three districts; New-England states, east of the Hudson and lake Champlain; middle states, whose territories extend west of the mountains or natural limit; and southern states, where slavery prevails; while the western states will soon be divided in three natural districts,—north of the Ohio, south of the Ohio, and west of the Mississippi, whose features and interests will also assume their own peculiarities, the presumable result of which will be a happy balance of indivisible interests.

We wish that a hint of Mr. Maclure's might meet the eyes of some of those who direct among us the education of youth. He insinuates that we may reasonably hope that, ere long, some portion of time will be appropriated, in our colleges and universities, to studies of evident utility, and that the knowledge of substances, their properties and their uses, will be permitted, in some degree, to encroach on the study of mere words, or the smattering of dead languages. His hopes begin to be partly realized, and the utility of the study of our soil, our waters, our minerals, our fossils, our plants, our animals, &c. is becoming daily more evident; let us hope that these studies will soon be taught every where, together, at least, with those of a less permanent and general utility. We shall conclude in the words of this author,—" The earth is every day moulding down into a form more capable of producing and increasing vegetable matter, the food of animals, and consequently progressing towards a state of amelioration and accumulation of those materials, of which the moderate and rational enjoyment constitutes great part of our comfort and happiness. On the surface of such an extensive and perpetual progression, let us hope that mankind will not, nay, cannot, remain stationary."

These remarks bear evidence that our worthy author is gifted with a philanthropic and philosophical mind. The style and the details of his work bear the stamp of the same modest, unassuming, and plain philosophy, and give the author a title to the highest reward of a good citizen, the gratitude of his countrymen; and should his labours be rewarded with the praise that greeted his predecessor Volney, we doubt not he will feel his anticipations fully realized. C. S. R.

REVIEW OF An Index to the Geology of the Northern States, with a Transverse Section from the Catskill Mountains to the Atlantic, by Amos Eaton

C. S. Rafinesque

THE modern science of geology has already acquired teachers and students in our own country; it is deemed an essential branch of physical knowledge in Germany, France, Italy, Great Britain, &c. and within a short period, a desire appears to prevail with us to keep pace with them, at least in the knowledge of our own soil. Since the general views of Volney and Maclure were published, many local labours have appeared, among which those of Dr. S. L. Mitchill and Dr. Drake, deserve an exalted station; and now, we have, in the attempt of Mr. Eaton to elucidate the geology of Massachussets, &c. the results of more than 1000 miles of travels on foot, the real way to observe with attention, and survey minutely. We were acquainted with Mr. E. already as a competent botanist, and he now introduces himself before the public as an attentive geologist. We shall follow him with pleasure in his new capacity, being thoroughly convinced that it is merely by such accurate observations and zealous exertions, that the science of practical geology can be successfully cultivated, and attain all the certainty of which it is capable. When accurate observers will spread themselves all over our states, and communicate the result of their researches, the practical benefits likely to arise therefrom, will be more generally felt; then, and only then, general geologists will become enabled to draw true conclusions, and frame lucid theories.

This remark is enforced upon us by the tract which we have undertaken to examine, and which somewhat invalidates the preposterous conclusions of Mr. Maclure, when he asserted that all the New-England states were of primitive formation. Mr. E. has been enabled to ascertain that nearly 18 different varieties of formations exist between Boston and the Catskill mountains, including nearly all the classes of formations. If Mr. Maclure meant to tell us that the primitive formation of granite, gneiss, slate, soapstone, &c. were prevalent in those states, at the surface or at a certain depth, he might perhaps be correct in his assertion, although we might ask him if he doubts that such a formation exists almost every where at a particular depth?

The successive formations through Massachussetts, &c. appear to underlay each other in the following order of strata, beginning from the surface; 1. Alluvial; 2. Basalt; 3. Rocksalt; 4. Gypsum; 5. Compact limestone; 6. Breccia; 7. Red sandstone; 8. Rubblestone; 9. Graywacke slate; 10. Argillaceous and silicious slate; 11. Metalliferous limestone; 12. Sienite; 13. Calcarious and granular quartz; 14. Soapstone; 15. Micaslate; 16. Gneiss; 17. Granular limestone and quartz; 18. Granite. But they are not always superincumbent on each other, although they never deviate from this numeric alternative stratification, even when many stratas are missing; the granite appears on the surface of the soil near Hinsdale, Chesterfield and Spencer, while it is covered with one or more of the above stratifications every where else. Mr. Eaton has come at this result, by an attentive observation of the successive appearance, and nature of the immediate strata under the soil, in an alternating progress. When, for instance, he has found several successive stratas in the eastern part of the valley of the Connecticut, and then finds them again in an opposite order, west of the river in the same valley, he is led to believe, with the greatest degree of probability, that they extend under the river in a proportionate succession and depth. But we regret that, led too far by the happy result of this discovery, he is induced to suppose, that the strata found east of the Hudson, are carried under it, and under the *Catskillmountains*, although he has not observed their re-appearance beyond them. This amounts, at the most, to a plausible hypothesis, but not even to a probable theory, if we consider that this supposition requires that those strata should extend under our whole continent, as far, perhaps, as the stony mountains in the west; while those fronting the Atlantic, ought to sink under the ocean, and appear again in Europe in the same order, which is not the fact.

This proves the danger of systematising and speculating on insulated facts, what is true in the valley of the Connecticut and near Worcester, must not be extended on either side to Europe and Asia; it is very possible, and even very probable, that many strata belong to local or limited formations, wherefore they may disappear when we should the least expect it. No formation ought to be considered as universal and continued, except the granitic; and some doubts may be entertained as to the truth of this supposition, notwithstanding the observations that are deemed conclusive.

The value of this pamphlet, does not, however, depend upon the occasional theories assumed, but upon the multitude of local facts, and the attentive study of the soil, in a progression from east to west. The observations of the author deserve to be read and considered by all those who deem a knowledge of our soil important, and they throw much light upon the whole geology of New-England; and even New-York. It appears that nearly one-third of the surface of this section, is composed of an alluvial soil, part of which is river alluvial.

We consider the whole as a good attempt towards the requisite knowledge of the surface of the soil, in the region observed, certainly a better one than Mr. Maclure's in its local capacity; but we presume that many other perambulations and excursions, and much research, are yet requisite, before a complete idea of the soil of New-England can be formed. A section of a base from New-York to Cape Cod, and another across the White Mountains, would be particularly desirable.

According to the remark of our author, a geological section of a country must always be rather a caricature of it, than a correct delineation: if we were to consider in that light his geological section, we should call it a very clever hypothetical caricature: it is, however, preferable to Mr. Maclure's sections of the United States, although both are defective in a different light; this last by carrying the formations perpendicular, as if they radiated from the centre of the earth; while Mr. Eaton's section shows the undulations and progressions of the strata, but often makes them reach a depth to which they are perhaps unknown, or gives them an extension, to which we have no proof that they reach. He divides the different strata, of which the soil of New-England is composed, into five classes, primitive, transition, secondary, superincumbent, and alluvial. We shall say a few words on the second and fourth of his classes. The transition formation, which is borrowed from Werner, is totally illusive in name and application: when transition rocks are crystallized in mass, they belong to the primitive or crystallized formation; when they are deposited in thin layers, or thick continued strata, they belong to the secondary, or deposited formation; when they are composed of agglomerated fragments, they belong to a subdivision of the same formation which may bear the name of agglomerated. The name of superincumbent rocks is given to the basalt, greenstone, trap and amygdaloid rocks, which belong to the volcanic or emitted formation. We must observe that he is mistaken, when he gives the following definition of volcanic productions, viz. " minerals upon which changes have been wrought by volcanic fires." Since the luminous discoveries of Patrin and Davy on volcanic productions, they must be termed, *minerals chemically emitted and combined*. The emission of water, mud, &c. by igneous volcanoes, the aerial volcanoes or volcanic springs, existing every where, and emitting air, clay, sulphur, hydrogen, &c. with or without heat and fire, the numberless submarine volcanoes, yet existing under the sea, and forming there, when compressed by a great weight of water, stratas of basalt, trap, coal, &c. by means of their smoke, ashes and fluids, are evident proofs of the emitted or volcanic origin of many of the secondary formations; and it would be difficult to prove that all those secondary substances which cannot be held in dissolution in air or water, or formed chemically in the sea and the atmosphere, do not belong to the same volcanic formation.

We shall not attempt to confute the absurd supposition that the strata, now constituting the Catskill Mountains, and the western parts of New-York, once extended to the Atlantic ocean. This speculative hypothesis, ought at least, to be supported by very strong proofs before it is advanced, and we are unacquainted with the power that could remove this chain of mountains, without disturbing the regularity of stratification, upon which this hypothesis is built; while we know very well that similar local causes may produce here and there, detached masses of consimilar substances.

The chain of mountains which divide the waters of the Hudson from those of the Connecticut, are called the *Peru* Mountains by Mr. Eaton; we thought hitherto, that their name was the *Tackonick* Mountains, while the *Peru* Mountains are a chain in the state of New-York, west of lake Champlain, where the Hudson takes its rise; we refer those, who may have any doubt on the subject, to Spafford's Gazetteer of New-York, and beg leave to ask who is in the wrong, Mr. Spafford or Mr. Eaton?

We regret that the premature geological speculations of Mr. Eaton, should have induced him to add to his valuable details of facts, an appendix under the title of *Conjectures respecting the Formation of the Earth*. It is in reality the common, but deplorable propensity of all geological writers, to deduce and assume some theoretical hypothesis, as soon as they have observed or collected a few facts, changing thereby gealogy into geogony, which are two different sciences altogether. The former describes the earth as it is, and no one will venture to deny its conclusions, since they arise from facts and existing causes, while geogony describes the earth as it was, or rather as it is supposed to have been, at different periods, or attempting still more, ventures to assert what it may yet become; when the speculations of geogony are deduced from history, records, data, remains, analogies, and phenomena, they become a sort of geological history; but all those which emanate from suppositions, conjectures, fictions, presumptions, probabilities and plausible causes, are at best but ingenious dreams, particularly when they attempt to embrace the origin and the end of our globe. Such are in part, the features of the conjectures before us: being not even modelled from the actual knowledge of the various parts of the globe, neglecting more or less the enlarged views, which late discoveries have revealed, the immense strata and mountains of organic formation scattered every where, and even under other formations, the various volcanic formations covering one third of the known soil, the numberless anomalies through the strata, their different succession, arrangement and configuration in different parts, and a variety of other important considerations; and they speak, instead of a primordial chaotic mortar, of an internal heat of the earth lifting up the granite, of an antediluvian continent, which has sunk and disappeared, &c. mere conjectures indeed, since they may be so easily denominated, when we attend to the actual phenomena and formations going on before our eyes. In the present improved state of chemical knowledge, from which our age has received the appellation of the age of chemical philosophy, every former conjectural theory must shrink before the chemical theory of the formation of the earth, until another improvement of philosophical knowledge, or till new discoveries shall compel us to lay it aside, for something apparently better, or nearer to truth, according as our perceptions shall permit us to conceive it.

However, when Mr. E. states physical

or historical data, such as the deviation of the pendulum, the progressive succession of organized beings, the late comparative period of human existence, &c. we find him in the true line of logical geogony. When he attempts to show that the geogony of Moses and his account of the flood, do not in the least contradict the facts which experience has revealed, when he proves that the days of the creation have been periods of time, as many learned divines have asserted, and every geogonist believes; we find him engaged in a desirable act of conciliation between science and religion; which, those who may happen to be acquainted with the late radical Hebrew translation of the first chapters of Genesis, by the learned Olivet, may improve into a demonstration, against those who hold the doctrine of their literal translation and explanation. The prejudices which ignorance or sectarian tenets, had thrown over geological studies, as soon as they became involved or blended with geogony, may thereby, we trust, subside entirely; their removal is certainly desirable, and cannot fail to become acceptable to all the friends of mental union and peace.
C. S. R.

AGENDA, OR SELECTION OF QUERIES

H. H. Hayden

As facts are essentially important in all researches instituted for the promotion of science; the following AGENDA, or SELECTION OF QUERIES, is respectfully recommended to the attention of Geologists, Mineralogists, and other persons of correct observation, as being intimately connected with the subjects contained in this work, and calculated to aid and assist in all future researches of a similar kind.

OF MOUNTAINS.

1st. What is their mean height, and what their course or direction?

2d. Is one side of a range of mountains, or hills, more abrupt, steep, and broken, than the other? If so, which side is it, "and to what point of the compass is it opposed?"*

* Question by the London Geological Society.

3d. If a steep and craggy appearance present itself on one side of a mountain, does it face an extensive valley through which runs a river, large or small? If so, what is the direction of its current, and the greatest height to which such river has been known to rise, from rains, melting of snows, &c.?

4th. If such river exist, is its *general* course through the middle of the valley, or does it run any distance near the foot of the mountain?

5th. If a river pass either obliquely or at right angles through the mountain, where rocks are presented to view, to any considerable height above the water, are there any appearances of the operations of currents upon the rocks, above the greatest height that such river has been known to rise? If so, what appears to have been the direction of such currents? This may be easily determined by the following remarks:

1st. The parts, or points of rocks, against which the currents were opposed, will present a smoother surface and more worn than the side which looks down the stream, or against which the current was not opposed.

2dly. Pot-like holes, formed in the rocks by the operation of currents, are often observable in situations far above the present level of any streams in their vicinity. A careful examination of these, will enable the observer to determine the course of the current by which they were formed—and by the following marks:—The side over which the current flows into the hole, is generally shelving under, on the up-stream side. This is occasioned in the following manner: When the current is propelled into the pot-like hole, the pebbles which are already within it, are driven with considerable force against the up-stream side. If they occasionally fall into the current, when it strikes against the lower or down-stream side, they are forcibly thrown back again, and thus kept playing against the upper side, by which means the hole becomes shelving under. Lastly, the sand and pebbles that are occasionally driven out, produce, by abrasion upon the down-stream side, an ewer-like process or gutter, which is very perceptible in many of them.

6th. If there are any appearances of a part or portion of the side of a mountain having slidden down to its base, what appears to have been the most probable cause of its removal? and what its *original* height above the mountain's base?—And moreover, is there any narrow, but extensive valley or channel through subordinate hills, and through which a current may have run, directed against a point where such portion of earth, or rocks, have slidden off? If so, what is its breadth, extent, and direction, in relation to the ridge, or range of the mountain?

7th. Are there any vallies, or gaps, that intersect a range of mountains, either strait or circuitous? and what is the greatest probable height of the highest point in such valley or gap, above the mountain's base?

8th. Are there any appearances of the operations of currents in such vallies, either in the earth or upon the rocks that may be exposed; and above the height at which the streams that at present flow in them? If so, at what height above the bottom of the vallies do they run, and what appears to have been the direction of such current?

9th. Are there any considerable quantities of alluvial grounds, where the waters of such vallies are discharged into the vallies adjacent to the mountains? and have they increased materially within the memory of man? If so, what is their extent, and probable increase, annually?

10th. Are there to be found on the tops or sides of mountains, composed principally of granite, or primitive limestone, detached masses of transition, or secondary rocks, out of place? If so, on which side of the mountain are they found, and at what height above its base? And also, of what size, description, or character, are such masses; at what distance are such rocks found in place, and in what direction from the masses so found?

The same remarks are applicable to mountains, the rocks of which are of a different order, (viz.)

11th. Are there to be found on transition mountains, masses of secondary rocks? on which side, and at what height? At what distance, and in what direction, are rocks of the same kind found in place?

12th. On a range of mountains of secondary formation, are there to be found masses of primitive rocks such as granite, or those of transition? &c. &c.

OF EXTENSIVE, BROAD, OR NARROW VALLIES, OR IN-
TERVALS.

1st. Are they *generally* level, or broken and interrupted? If of the former kind, do they appear to be composed in any degree of alluvion? This may be determined by several means. 1st. By the nature of the soil, being either of sand, or gravel, and differing materially from that at the bases, and on the sides of adjacent mountains. 2dly. By ditching, canaling, sinking wells and other works, by which the structure and character of the earth beneath the surface is exposed to view. If composed of alternate layers of sand, clay, and pebbles, in horizontal, inclined, or undulating strata, with occasional deposites of fossil wood, or organick remains, it may reasonably be considered as alluvial. If on the contrary, it is of an uniform texture, and presenting none of the above marks, it may be considered as not having been disturbed, and as original.

2d. If the earth, thus exposed, appears stratafied and horizontal, what is the order in which they occur, and to what depth has this appearance been known to extend?

3d. If the strata are inclined and undulating, or wave-like, what is the dip or inclination of such strata, and what the *general* direction of their dip?

4th. If pebbles occur in such strata, are they in nids, or nests like, or uniformly distributed to any considerable extent? Are they rolled, or rounded, and of what size?

5th. Are they uniformly of one kind in substance, or different; and what are the kinds?

6th. Are there rocks of a similar substance, in place, at any distance from them, and what is the distance and direction?

7th. If the vallies are broken and interrupted, is it by spurs of mountains, ridges of rocks, or isolated hills, composed in a greater or less degree of pebbles?

8th. If of spurs of mountains extending to any distance into the valley, are there any appearances on either side, of the operations of currents which may have been opposed to, or set against them? If so, what are the appearances, and on which side of the spurs in relation to the compass, are they?

9th. If interrupted by ridges of rocks, of what kind are they, and in what direction do they run in relation to the valley, and do they discover any marks of abrasion by the operation of running water, &c.?

10th. If interrupted by isolated hills, composed mostly, or in part, of rolled pebbles, what is their character, and are there rocks of the same substance in place, in any direction from such collection of pebbles? If so, what is the direction from such hills, and at what distance?

11th. If such accumulation of pebbles are to be found in a valley, are there any deep and extensive ravines, vallies, or gaps, in the neighbouring mountains, or hills, through which a violent current may have ran, and by which such pebbles may have been transported to some distance in the low grounds, where they are found? If so, in what direction are the vallies from the pebbles thus collected, and at what distance?

11th. If it is reasonable to suppose that the pebbles so collected, may have been transported from the mountains by a current, or currents, *and they are thrown up into hills,* or small eminences; it is a fair conclusion, that the currents by which they were transported, were checked or opposed in their course, by opposite or lateral currents from other directions. To determine this, it is necessary to examine whether there are any deep vallies or gaps, through the mountains or hills, on the opposite side of the valley, where such pebbles are found, and through which currents may have flowed in a direction opposite to the first.

This fact may be often observed in the high and sudden rise of waters, by heavy rains, melting of snows, &c. In order, however, to obtain an accurate view, and correct information on this point, it is necessary to examine the subject from an elevated situation; as on the side or top of a mountain, from which the eye can take in at one view, all the narrowings and widenings of the great valley; the sinuosities of the mountains; and the cross cuts or gaps of the mountains, that open into the great valley or vallies.

12th. If from such a view, there should be found vallies, or gaps, through the opposite range of mountains, what is their breadth, extent, course, and direction, in relation to the accumulated masses of pebbles, or other substances differing from the common earth?

13th. If transverse vallies occur (so called by *Saussure*, in contradistinction to longitudinal vallies, which are such as have an extensive range between two parallel ridges or mountains,) are there any appearances of the operations of currents from the lateral vallies; such as hillocks of sand, pebbles, boulders, or rocks, at or near the junction of the transverse with the longitudinal valley? If so, in what direction do the transverse vallies run, and on which hand do the deposites appear; whether on the right or left?

14th. " If the lateral valleys, which terminate at a principal valley, as the branches of a tree at its trunk, correspond or not; or, in other words, whether the branches of that trunk are opposite or alternate?"— *Saussure.*

" The answers to these two questions are very important, for the solution of this question : whether the valleys have been excavated by currents of the sea?" —*Saussure.*

15th. If a valley, on one side of a range of mountains, appears to be underlayed with rocks, primitive or secondary; stratified or unstratified; horizontal or inclined, are the rocks, if any, in the valley on the opposite side of the mountain or range, of the same kind, and arranged in the same order, so as to afford any reason to believe that they underlay the mountain? This is a question of no small importance to the geologist.

16th. If a valley occurs, of great or small extent, surrounded by hills or mountains, is there reason to believe that it was ever a lake? If so, what are the reasons?

17th. If it is supposed that in such a situation, a lake ever existed, what are the most probable means by which it was filled up?

18th. Is there one, or more, natural openings through the surrounding hills or mountains, at a small elevation above the present level of the valley? If so, of what description are those openings, and in what direction in relation to such valley?

19th. If there are no openings of the above description, are there any appearances of a disruption of the hills or mountains, that surround such valley? If so, what are the appearances, and on which side, or sides of the valley are they?

20th. If there are appearances of a disruption of the hills &c. by water, (which must be supposed, if the lake was filled up with alluvion,) what is the most probable cause of its having been put in operation?

21st. Are there any appearances of a neighbouring valley, which may have, likewise, been a lake, and at a higher elevation than the first? If so, what are the appearances?

22d. In this valley, where a lake is supposed to have existed, are there any detached masses of rocks, either rounded or angular, buried in the earth, mixed with the soil, or distributed over the surface of the ground? If so, of what description are they, and in *what direction,* and at *what distance,* are rocks of the same kind found in place?

23d. In digging into the earth in such a situation, are there to be found organick remains of vegetables; or those of animals, either land or æquatick?

24th. If of vegetables, or fossil wood, of what description are they?

25th. If of land animals, are they of a species that are indigenous, foreign, or of such as are extinct?

26th. If of æquatick, are there analogous to be found living? If so, where are they to be found, and of what kinds are they?

27th. Are such remains found only at a certain depth? or are they distributed generally through the alluvial mass? If the former, at what depth are they found?

28th. In a situation where it is supposed, a lake once existed, is the surface of the valley generally level? or is there a gradual descent from one side, or end, to the other? If the latter, in what direction is the descent?

29th. What is the probable height of such a valley, or valleys, above the level of the ocean?

OF RIVERS.

1st. " What is the extent of their course, and their inclination from their sources to their mouth."—(*Saussure.*) Do they discharge their waters immediately into the sea, or bay; or into a gulf, or arm of a bay?

2d. Is the country through which they run, generally mountainous or hilly? or is it low and flat?

3d. What is the mean rate at which the current generally flows?

4th. What is the highest rate at which the current of a river flows, in any part of its course, for any considerable distance, during, or at nearly low water?

5th. Is the water of a river or rivers, clear and transparent to any considerable depth, during the winter and summer months?

6th. If the current of a river is rapid, for any distance, in those seasons, does there appear to be any alluvion mixed with the water; or sand moving upon the bottom?

7th. If the course of a river is between two ranges of mountains, with extensive meadows, or intervales on its borders, are they rocky or alluvial?

8th. If alluvial, what is the mean height of the banks?

9th. Are they steep and broken, or gradually descending towards the water?

10th. If high and broken, what are the appearances which they present?

11th. Are there any appearances of fossil wood, or other organick remains to be seen in the banks, and of what kind are they?

12th. If in such banks organick remains are to be seen, are they *at, or below low water mark,* or gradually distributed in the earth from the water to the top of the banks?

On many large rivers, and, in some instances, on smaller ones, there are two and sometimes three alluvial banks on each side, except where the river passes through a mountainous district. On this important subject, it may not be amiss to offer a few remarks previous to proposing any interrogatories.

The lands immediately bordering upon a river, and which form the first bank on each side, are generally considered as the intervales or meadows, and, if alluvial, have generally a gradual descent from the river for a half a mile, and from that to two miles or more, where there commences another bank or range of hills, likewise of alluvion, from twenty to sixty feet or more in height. This second tract of country or land, generally extends to the foot of the mountains, and varies in breadth from one to several miles.

These two, and sometimes three banks or tiers of alluvial land, have been distinctly mentioned by several travellers, as occurring on many rivers; but no particular description has hitherto been given of them, that I can find, neither have any remarks been offered that are calculated to make us acquainted with their history, or the cause of their formation. In a geological point of view, they are extremely interesting and important; so much so that, on a careful examination of the order of their arrangement, in relation to the river on which they lie, and the adjacent mountains or hills, and their internal structure, no one will hesitate to admit, that they distinctly point to *two important epochs* or events, that have taken place upon this globe, and by which they were probably formed. It is from this view, that I am induced to invite the attention of the naturalist more particularly to them, and to propose the following questions for his observance.

13th. Whenever two or three alluvial banks, of the above description occur, what is the course of the river through the entire extent of such district?

I mention "such district," because in some cases, a spur of a mountain, or ridge of rocks, crosses a valley, breaks off at a river, and interrupts the extension of those alluvial banks, and also occasions a bend or difference in the course of the river, below which the alluvial banks occur again, &c.

14th. What is the mean breadth of the first intervales, or alluvial banks next to the river, and what their height?

15th. What is the mean height and breadth of the second alluvial banks, on each side of the river?

16th. Do they ascend or descend towards the mountains or hills?

17th. Of what do those districts appear to be composed, at a small depth below the surface, and to the greatest depth to which they have been explored?

18th. Of the component parts, which is the most predominant; sand or clay, &c.?

19th. Do springs of water occur, at the foot of the second alluvial banks *generally?*

20th. Have mineral springs been known to occur in those districts, and what are their properties?

21st. Are they resorted to as such, and are they perpetual?

22d. In digging for wells, and other purposes, in the second alluvial plain, has an instance occurred in which it has been carried *below the level* of the lower, or first intervale on the river? If so, what were the appearances?

23d. Were there any appearances of fossil wood, or other organick remains to be seen, *particularly* on a level with the first bank or alluvial district?

24th. Are there to be found in the earth upon the second plain, irregular masses or boulders of rocks, out of place?

25th. If so, are they exclusively confined to the upper plain, or are they alike distributed in the earth, both in the first as well as second bank?

26th. Of what description are they, and in what direction, and at what distance are rocks of the same kind found *in place?*

27th. Have beds of salt or fresh water shells, or shell-marle, been found in either of those banks? If so, which bank is it, and of which kind of shells, and at what depth are they found?

28th. Are there rocks, shells, &c. to be found in the alluvial banks on both sides of a river, or only on one side? If the latter, which side is it?

29th. Are the alluvial banks or intervales on one side of a river, generally *of a greater breadth than those on the opposite side?* If so, on which side do those of the greatest breadth lie, and what is the difference from those on the opposite side?

30th. If a river has a southerly course, from any point, between north west and north east, and after running through a mountainous or rocky district, enters upon a district entirely alluvial, are there to be found in the latter district, masses of rocks and rolled pebbles of the same description as those through which the river passes above? If so, how far do they extend in the alluvial soil?

31st. Are the pebbles uniformly distributed through the soil, on both sides of a river, or are they more abundant on one side than the other?

32d. If, as before, a small river has a southerly course, through a rocky district of country, and in its descent, receives an auxiliary branch from an easterly direction, are not rolled pebbles more abundant, in the south and west bank of the principal stream, in the direction of the auxiliary branch?

33d. Wherever rolled pebbles prevail on the borders of rivers, creeks, &c.; are they not found in greater quantities on the banks on the south side, than in those on the north?

34th. Wherever pebbles are found, as above, do they not diminish in size, as we recede from the river or creek?

35th. Where auxiliary branches pass through a rocky district, and discharge their waters into a river on the east, or north and east side, are rolled pebbles found at their mouths in greater quantities than elsewhere? If so, are there like quantities found at the mouths of auxiliary branches, running through similar districts, and discharging themselves on the west, or north west side of a river?

36th. If a river has a southerly course, and an auxiliary branch falls into it on the west side, on which side of the branch are the alluvial banks the highest? and *in particular,* where there are two alluvial banks.

37th. If a river discharges itself into the sea, how high does the tide of the latter rise at that place?

38th. At what rate does the tide, at half flood, flow up the river, and to what extent does it check the current of the river?

39th. If there are deltas at the mouths of rivers, what are their lengths, breadths, and heights, above the level of the sea?

40th. Are they ever generally or partially overflown by the tides of the sea, or freshes of the rivers?

41st. What are the appearances of their banks?

42d. To what extent have they been known to increase, within the longest known period?

43d. Are they covered with forest trees? If so, of what kind and size are they, and how near do they approach the sea?

44th. Are there any sandy districts or deserts in the vicinity of such deltas? If so, how are they situated, and what are their lengths, breadths, &c.?

OF ALLUVIAL DISTRICTS AND PLAINS, WHETHER INCLINED, ON THE MARGIN OF LAKES, OR ON THE BORDERS OF THE OCEAN.

1st. "What their shape and extent, with the nature, height, and general appearance of the hills or mountains, by which they may be bounded." (*London Geological Society.*

2d. "The degree and direction of the inclination or slope?" (*Ibid.*)

3d. If plains, or alluvial districts, occur on the margin of a lake or lakes, on which side are they, and are their corresponding plains or districts on the opposite side, or in any other direction on the borders of the lake?

4th. If wells, canals, or other excavations, have been made in such districts, to what depth have they been carried, and what are the appearances that are presented to view?

5th. Are fossil trees found in digging in such districts?

6th. If so, at what depth below the surface do they occur, and are they found only at a certain depth, or occasionally distributed from the surface to the bottom of a well, canal, &c.?

7th. If trees are thus found in the earth, do they appear to be thrown together promiscuously; or do the tops appear to lie in one direction, as is the case at Yule, in Yorkshire, England?

8th. Are organick remains of animals found in such situations? If so, at what depth, and of what kind are they?

9th. Are there found, beneath the surface, in such districts, rolled or angular masses of rocks? If so, in what direction, and at what distance, does the same kind occur in place?

10th. Is it reasonable to suppose, that such districts could have been formed by the operations of any river that at present flows, or may have flown, in the vicinity of such districts? If so, what must have been its course?

THE END.

Part VIII
AMERICAN GEOLOGY COMES OF AGE

Editor's Comments on Papers 29 and 30

29 THE EDINBURGH REVIEW
Reviews of Observations on the Geology of the United States by William Maclure *and* An Elementary Treatise on Mineralogy and Geology by Parker Cleaveland

30 AMERICAN GEOLOGICAL SOCIETY
Constitution, Officers, and Proceedings for 1819 and 1820

The selections in this volume illustrate the growth of American geological thought. From a few articles and pamphlets on such isolated phenomena as earthquakes and fossil bones, geology advanced to embody a unified descriptive science based on a foundation of systematic rock and mineral classification. Scientists contributed to this development as researchers, collectors, educators, and editors of scientific periodicals. The contributions and encouragement of foreign visitors, the industry and generosity of mineral and fossil collectors, and a public eager to learn about their nation's geologic structure and past were also essential to this growth. The introduction of geology to the United States was indeed a complex process.

By 1820 American geology had come of age. Amos Eaton cited this year as the close of the first era in American geology, and subsequent historians of science have concurred with Eaton's appraisal (Merrill 1924; Greene 1969). The final two selections, originally published in 1819–1820, reflect the maturity of American earth science at the close of the Maclurean era.

American earth scientists owed a great debt to European geologists who pointed the way in earth studies. As evident in the original sources reprinted in this volume, American publications frequently contained acknowledgments to these foreign leaders. In 1819 it was America's turn to receive European praise, as the prestigious *Edinburgh Review* considered the merits of William Maclure's "Observations" and Parker Cleaveland's "Elementary treatise" (Paper 29). The anonymous Edinburgh critic termed both works "very excellent publications," and took special interest in the reports of North America's natural resources. The interest and enthusiasm of the British reviewer certainly outweighed the few criticisms, and American earth scientists pointed with pride to the attentions of their trans-Atlantic colleagues.

The formation of the American Geological Society, documented by the publication of the society's constitution, officers, and proceedings (Paper 30), was another indication of the enthusiasm and vitality of earth science in North America. Though based at Yale College in New Haven, the society had a membership that was truly national in scope; scientists from Maine, Massachusetts, New York, Connecticut, and Pennsylvania were among the first officers. William Maclure was elected president, a sign of his colleagues' admiration and esteem. In addition, the society provided for eight vice presidents. This seemingly excessive number is best understood in terms of Article VII of the society's constitution, which defines a quorum as one executive officer and four members. Eight vice presidents presumably assured a quorum at future meetings and also distributed honors among the society's founders. With such eminent geologists as Silliman, Cleaveland, Hitchcock, Eaton, Mitchill, Gibbs, and Rafinesque in its ranks, the American Geological Society served as a focus for the exchange of information and ideas and provided the first organized representation for professional geologists in North America.

In the next decades American geologists enjoyed steadily increasing prestige and responsibility, as both state and national governments commissioned scientific surveys of the vast continent. The growing quantity and quality of geological data inevitably led to specialization among earth scientists, and monographs on North American minerals, fossils, mining, and soils were published to more fully document the nature of the earth's crust. These factual compilations in turn enabled American geologists to formulate ever more sophisticated theories of the earth's structure and its history. Thanks to the efforts of the early scientists, educators, explorers, and collectors, there was laid a secure foundation on which American geology was to grow and prosper.

REFERENCES

Greene, J. C. 1969. The development of Mineralogy in Philadelphia, 1780–1820. *Am. Philos. Soc. Proc.* **113**:283–295.

Merrill, G. P. 1924. *The First One Hundred Years of American Geology*. New Haven, Conn.: Yale Univ. Press.

REVIEWS OF Observations on the Geology of the United States by William Maclure AND An Elementary Treatise on Mineralogy and Geology by Parker Cleaveland

[From the Edinburgh Review.]

IN a former number,* we gave an account of a new Mineralogical Journal, published in America, by Dr. Bruce of New York. We hailed the appearance of this work as a proof of the attention that had been excited to this interesting branch of science, in a field so sure to yield an abundant harvest; and it was with regret that we learned, that a journal which promised so well at its outset, had very soon been discontinued.

We have now great pleasure in introducing to the notice of our readers, two very excellent publications, which abundantly prove that the study of mineralogy is pursued with no less eagerness and success in the United States, than it has been for some years past in most of the countries of Europe. There is not perhaps any department of science which, at the present time, merits a greater degree of attention in that great and prosperous country, from its various practical applications to some of the most important sources of national wealth and power; and the more especially that, from the limited researches already made, nature appears to have added, in abundance, some of her most valuable mineral productions to the other internal resources which she has lavished in that part of the world.

The geological part of Mr. Maclure's book was first published in the sixth volume of the American Philosophical Transactions; in the present edition there are some additions and corrections,

* Vol. xvii. p. 114.

besides two new chapters, which the author informs us in his preface, are ' an attempt to apply geology to agriculture, in showing the probable effects the decomposition of the different classes of rocks may have on the nature and fertility of soils. It is the result of many observations made in Europe and America, and may perhaps be found more useful in the United States than in Europe, as more of the land is in a state of nature not yet changed by the industry of man.'

Mr. Maclure appears to be very thoroughly conversant with his subject, and to have studied with great attention the geological structure of a considerable part of Europe. He is a disciple of Werner; but we recognise him as such, more by the descriptive language he employs, than by his theoretical opinions. His general views are much more enlarged and philosophical, than is usually met with in the geologist of that school; and, like most of those who have had opportunities of extensive observation, he has found that the theory of the Freyberg professor is of a very limited application. The following remarks in his preface are a sufficient proof that his geological creed is not that of Werner.

' In all speculations on the origin, or agents that have produced the changes on this globe, it is probable that we ought to keep within the boundaries of the probable effects resulting from the regular operations of the great laws of nature, which our experience and observation have brought within the sphere of our knowledge. When we overleap those limits, and suppose a total change in nature's laws, we embark on the sea of uncertainty, where one conjecture is perhaps as probable as another; for none of them can have any support, or derive any authority from the practical facts wherewith our experience has brought us acquainted.'

While we acknowledge the valuable information which this little work conveys, we cannot bestow any praise on the manner in which the materials are put together. There is a great want of method and arrangement; for, although the author has laid down a very good plan, he has not adhered to it, but has mixed up one part of his subject with another, so as to cause considerable confusion; and, were it not for the accompanying coloured map, it would often be very difficult to comprehend his descriptions.

The Elementary Treatise of Mr. Cleaveland, is a work of considerable merit. He has derived his materials, as he informs us, chiefly from the works of Hauy, Brochant, Brongniart, Lucas, Kirwan, and Jameson; but he has adopted Brongniart as his model; and in doing so, we think he has followed the most judicious and most useful of all the mineralogical writers who have preceded him. We entirely concur in the following remarks on the treatise of Brongniart, by the author in his preface.

' Many of the writers of the French and German schools appear to have indulged an undue attachment to their favourite and peculiar system, and have hereby been prevented from receiving mutual benefit; the one being unwilling to adopt what is really excellent in the other. But it is believed, that the more valuable parts of the two systems may be incorporated, or, in other words, that the peculiar descriptive language of the one may, in a certain degree, be united to the accurate and scientific arrangement of the other. This union of descriptive language and scientific arrangement has been effected with good success by Brongniart, in his System of Mineralogy—an elementary work, which seems better adapted both to interest and instruct, than any which has hitherto appeared.'

Although this book is necessarily compiled, in a great degree, from the writings of others, it contains much valuable information respecting the mineral productions of the United States. It is to this part of the work that we shall confine our remarks; and we feel disposed, for the sake of our general readers, to dwell chiefly on the information Mr. Cleaveland conveys, respecting those mineral substances that are connected with the advancement of that active and enterprising people in wealth and political importance, rather than upon the rarer productions, which are only interesting to the mineralogist.

There is one merit of Mr. Cleaveland's book that ought not to pass unnoticed; we mean the form in which it is published. It is printed upon excellent paper, with a neat and perfectly distinct small type; and the same matter is contained in one volume, which in England, would have been scattered over the surface of three. We should be glad to see it reprinted exactly upon the plan of the original; and we have no doubt that it would be found the most useful work on mineralogy in our language.

Coal exists in several parts of the United States in great abundance. We have already spoken of the vast series of coal strata westward of the Alleghany range, and of an extensive coal formation near Richmond in Virginia. In Pennsylvania, it is found on the west branch of the Susquehannah; in various places west of that branch; also on the Juniata, and on the waters of the Alleghany, and Monongahela. In Connecticut, a coal formation, commencing at Newhaven, crosses Connecticut river at Middleton, and embracing a width of several miles on each side of the river, extends to some distance above Northampton, in Massachusetts. There are also indications of coal in the states of New York and New Jersey. In Rhode Island, anthracite is found, accompanied by argillaceous sand-stone, shale with vegetable impressions, &c. similar to the usual series of coal strata. The coal at Middleton, in Connecticut, is accompanied by a shale which is highly bituminous, and burns with a bright flame.

' It abounds with very distinct and perfect impressions of fish, sometimes a foot or two in length; the head, fins, and scales being perfectly distinguishable. A single specimen sometimes presents parts of three or four fish, lying in different directions, and between different layers. The fish are sometimes contorted, and almost doubled. Their colour, sometimes gray, is usually black; and the fins and scales appear to be converted into coal. The same shale contains impressions of vegetables, sometimes converted into pyrites.'

Neither Mr. Cleaveland nor Mr. Maclure give us any information respecting the extent to which the coal has been wrought in any of the numerous places where it has been found, or the thickness of the seams. A scarcity of wood for fuel must be felt before coal will be sought after with much spirit; and there is probably still wanting in the United States that profusion of capital which can be risked in the uncertain operations of mining.

Iron is found in the United States in a great variety of forms, and is worked to a considerable extent. In the year 1810, there were five hundred and thirty furnaces, forges, and bloomeries, in the United States, sixty-nine of which were in the state of New York; and the iron manufactured at Ancram, New York, is said to be superior, for many purposes, to the Russian and Swedish iron. It is made from a hematitic brown oxide. Mr. Maclure informs us, that there is a bed of magnetic iron ore, from eight to twelve feet thick, wrought in Franconia, near the White Hills, New Hampshire; that there is a similar bed in the direction of the stratification, six miles north-east of Philipstown on the Hudson river; and, still following the direction of the stratification, that the same ore occupies a bed nearly of the same thickness at Ringwood, Mount Pleasant, and Suckusanny, in New Jersey; losing itself, as it approaches the end of the primitive ridge, near Blackwater—a range of nearly three hundred miles. This immense deposite of iron ore is contained in gneiss, and is accompanied by garnet, epidote, and hornblende. In the state of New York, magnetic iron ore is found in immense quantities on the west side of lake Champlain, in granite mountains. The ore is in beds, from one to twenty feet in thickness, and generally unmixed with foreign substances: large beds of this ore extend, with little interruption, from Canada to the neighbourhood of New York. Clay ironstone is met with in considerable quantities. In Maryland, there are extensive beds of it three miles S.W. of Baltimore, composed of nodules, formed by concentric layers. Bog iron ore occurs in such abundance in many places, as to be smelted to a great extent.

Copper in the native state, and most of its ores, have been found in different parts of the United States; but there are no mines of this metal except in New Jersey, and these do not appear to be worked with much success.

Lead has been discovered in a great variety of forms; and there are several extensive mines of it. In Upper Louisiana, at St. Genevieve, on the western bank of the Mississippi, there are about ten mines. The ore, which is a sulphuret, is found in detached masses of from one to five hundred pounds, in alluvial deposites of gravel and clay, immediately under the soil; and sometimes in veins or beds, in limestone. One of the mines produces annually about 245 tons of ore, yielding 66 2-3 per cent. There are mines also at Perkiomen, in Pennsylvania, 24 miles from Philadelphia.

The ore is chiefly a sulphuret; but it is accompanied by the carbonate, phosphate, and molybdate. In Massachusetts, there is a vein of galena, traversing primitive rocks, six or eight feet wide, and extending twenty miles from Montgomery to Hatfield. The ore affords from 50 to 60 per cent. of lead.

Gold has only been found in North Carolina. It occurs in grains or small masses, in alluvial earths, and chiefly in the gravelly beds of brooks, in the dry season; and one mass was found weighing 28 lib. In 1810, upwards of 1340 ounces of this gold, equal in value to 24,689 dollars, had been received at the mint of the United States.

Native silver, in small quantities, is met with at different places, but in no other form. Mercury and tin have not been found. Cobalt occurs near Middleton, in Connecticut; and a mine of it was at one time worked. Manganese and antimony are found in several situations. Sulphuret of zinc is found in considerable quantity in Maryland, Pennsylvania, New Jersey, and Massachusetts. In New Jersey, a new variety of this metal has been discovered, in such abundance, that it promises to be a very valuable acquisition to the United States. It is a red oxide, composed, of zinc 76, oxigen 16, oxides of manganese and iron, 8. It is reduced without difficulty to the metallic state.

The chromate of iron, both crystallized and amorphous, occurs in different situations; particularly near Baltimore, and at Hoboken, in New Jersey. This mineral is employed to furnish the chromic acid, which, when united with the oxide of lead, forms chromate of lead—a very beautiful yellow pigment, of which there is a manufactory at Philadelphia. It is sold under the name of chromic yellow, and is employed for painting furniture, carriages, &c.

CONSTITUTION, OFFICERS, AND PROCEEDINGS FOR 1819 AND 1820

1. *American Geological Society.*

AT the conclusion of our last number, we announced the formation of an American Geological Society and the passage of an act of Incorporation by the Legislature of Connecticut, conferring the necessary powers.

Agreeably to that act a number of gentlemen from different States, held a meeting on the morning of Sept. 6th, in the Philosophical Room of Yale College for the purpose of organizing the society.

Col. George Gibbs was called to the chair, and the plan of a constitution was laid before the meeting by a committee.

On the evening of the 7th, it was adopted, after undergoing various amendments.

A copy is subjoined with a list of the officers elected for the ensuing year.

Constitution of the American Geological Society.

ART. I.—There shall be a President, eight Vice-Presidents, one Recording Secretary, three Corresponding Secretaries, a Curator and Treasurer, a Committee of Nomination, and a Committee of publication; all of whom shall be annually elected.

Art. II.—The Society shall consist of not more than one hundred members :—and of not more than twenty-five honorary, and forty corresponding members.

Art. III.—Candidates for admission into the Society, must be proposed by the Committee of Nomination, and be chosen by three-fourths of the members present.

Art. IV.—The annual meeting of the Society, for the election of Officers, shall he held on the Tuesday preceding the second Wednesday of September, at such hour and place as shall be agreed upon from time to time.

Art. V.—The other stated meetings shall be on the first Mondays in December, March and June : and all the meetings may be adjourned by the Chairman, for not more than seven days from the dates above mentioned.

Art. VI.—Special meetings may be convened by resolution of the Society, or by public notice from the President ; or, in case of his absence, from the acting Vice-President, which meetings shall be restricted to the special objects of the Society, without power to enact regulations, or admit members.

Art. VII.—Five members, including the President, or one of the Vice-Presidents, shall form a quorum.

Art. VIII.—Every member shall pay to the Treasurer an initiation fee of five dollars, and shall be subject to an annual payment of one dollar.

Art. IX.—The Treasurer shall pay no money from the Treasury of the Society without a vote for this purpose and an order signed by the presiding officer.

Art. X.—The Society shall be located, provisionally, at New-Haven.

Art. XI.—No alteration shall be made in the Constitution, unless it be proposed in writing, at one of the stated meetings, previous to the annual meeting in September, and shall be decided by a majority of two thirds of the members present, at the said annual meeting.

Art. XII.—In such points of order as are not noticed in this Constitution, the Society will conform to the established customs of other similar institutions.

Officers.

William Maclure, President.	T. D. Porter, Curator.
George Gibbs, ⎫	A. M. Fisher, Treasurer.
Benjamin Silliman, ⎪	B. Silliman, ⎫ Committee
Parker Cleaveland, ⎬ Vice-	G. Gibbs, ⎪ of Nom-
Stephen Elliott, ⎪ Presidents.	P. Cleaveland, ⎬ ination.
Robert Gilmor, Jr. ⎪	R. Hare, ⎭
Samuel Brown, ⎪	George Gibbs, ⎫ Committee
Robert Hare, ⎪	J. W. Webster, ⎬ of Pub-
[Vacant.] ⎭	James Pierce, ⎭ lication.
T. Dwight, Porter, Rec. Sec.	

J. W. Webster, ⎫
F. C. Schaeffer, ⎬ Corresponding Secretaries.
E. Hitchcock, ⎭

The stated meeting for December having been postponed, a special meeting was held on the 26th of January, 1820, in the new Cabinet of Yale College.

Col. Gibbs, as first Vice President, took the chair.

Professor Silliman presented a memoir of considerable extent on parts of the counties of New-Haven and Litchfield, in Connecticut. He gave a connected view of the strata and formations from the old red sand stone, the green stone trap, and alluvial of New-Haven, through the succeeding clay slate, chlorite slate, and micaceous slate, to the Gneiss and Granite of the Alpine region of Litchfield county.

The extensive beds of white granular marble which alternate many times with the mica slate and gneiss of Litchfield county, and afford inexhaustible materials for architecture and the arts, were particularly noticed, as were the fine iron ore beds of Salisbury and Kent, and the spathic iron of Roxbury all of which are also situated in the gneiss and mica slate. Tremolite, garnet, staurotide, sappar, plumose mica stalactitical brown iron and graphic granite, occurred in fine specimens, in the tract described, and specimens of most of these were presented for the cabinet of the Society.

The same gentleman presented specimens of massive fluor spar, recently discovered in the parish of New-Stratford, town of Huntington, Connecticut, by Mr. Ephraim Lane, four miles south of his mine which affords bismuth, tungsten,* &c. According to Mr. Lane, this vein is two feet in width; and its immediate walls are white granular limestone which forms an extensive bed in gneiss. This fluor spar appears at two places, distant a fourth of a mile, and, when the snow is gone, will probably be found to form a gigantic vein. It has been observed only since the snows fell, and was first noticed in some fragments of lime stone, which had been quarried for burning.

The vein is much penetrated by quartz, mica, feldspar, and talc, but, it has been hitherto examined only on the surface. It is principally massive and its structure foliated or coarsely granular, but it presents well defined cubical crystals. Its colours vary from white to deep violet and purple, and are, principally various shades of the two latter. But the most interesting circumstance relating to it is its splendid phosphorescence. The light emitted when, it is thrown, in a dark place, upon a hot shovel, *is the purest emerald green;* pieces of an inch in diameter become in a few seconds, fully illuminated, and the light is so strong and enduring, that when carried into a room lighted by candles, or, by the diffuse (not direct) light of the sun they still continue distinctly luminous and the light dies away very gradually as the mineral cools. This interesting property was exhibited to the members of the society. Is not this variety of fluor spar *then the true chlorophane of Siberia?*

Prof. S. presented to the Society specimens of the green serpentine marble found near New-Haven, and which, according to the opinion of Mr. Brongniart of Paris, is the verd antique marble.

Col. Gibbs presented Smith's Geological Map of England, and various geological specimens; among which were varieties of the granite rocks of Haddam, Connecticut. These rocks contain tourmaline, garnets, sometimes of very great size—beryls and crysoberyl, both massive and crystalized. This being the only locality known in which the crysoberyl occurs in place, the specimens are therefore very interesting.

* See Vol. I, page 316 of this Journal

Mr. T. D. Porter presented some of the finest crystals of red oxid of titanium that have been any where found; the following memorandum accompanied them.

This titanium which I discovered in 1818, exists very well crystalized, and in comparative abundance, in masses of quartz which are scattered over the surface throughout the counties of Amherst, Campbell and Bedford, about twenty miles above Richmond in Virginia. Probably also it may be found in other counties contiguous to these, as the same rocks occur very extensively in all that quarter of the state; but I never had an opportunity to make any examination except in those I have mentioned.

Many of the specimens which I procured are superior both in size and in beauty, to any of the same species in the Cabinets which I have seen. A fragment of one crystal which I obtained, measures $1\frac{1}{10}$ in chesin diameter, and others are nearly as large. I have one specimen $3\frac{9}{10}$ inches in length, and another more than $3\frac{1}{4}$: both these are mutilated—the latter is broken off at each end, and was probably much larger; it is of the size of one's finger. The larger specimens are very liable to be thus injured, being exceedingly brittle. Their fracture is *commonly* foliated longitudinally and vitreous in the other direction. Frequently they are completely penetrated by quartz in the same manner as the green tourmaline of Massachusetts is by the Rubellite.

Like the different varieties of schorl, the greater part of the crystals were so compressed and striated, that their figure was very variable, oftener nearly cylindrical than of any regular prismatic form. I met with two or three specimens which were four sided prisms, truncated on each of the angles, having their terminations broken off and with a single crystal of four sides, which like those of the specimens just mentioned, seemed to meet at *right* angles and terminated very handsomely by a pyramid, whose sides corresponded with those of the prism. Many examples of crystalline termination were observed, but generally they were exceedingly irregular; sometimes one of the terminal planes was so large as almost entirely to obliterate the remainder. I believe I saw but two crystals with both ends perfect, among more than a hundred specimens which I collected.

A large proportion of the titanium found here, exhibited that peculiarity of configuration which is so characteristic of this mineral termed *geniculation*. In some cases a crystal was bent at but one angle; in others at many—and in others still, while the specimen was perfectly straight and smooth on one side, the opposite was marked by many flexures, a part only of the molecules, having apparently been subject to the law that determined to this angular form. Two specimens fell in my way which had all the angles rounded, appearing as if they had suffered partial fusion. Instances of the reticulated variety were rare.

The colour of the oxid, when taken from the interior of the masses in which it was imbedded, was a beautiful red, often accompanied by translucency; but that more exposed to the weather was commonly opake and almost black.

P. S. The Virginia titanium, although infusible by the common blow pipe, melts under the flame of the compound blow pipe of Prof. Hare, but is not reduced to the metallic state.

The Society directed Cases to be procured to receive specimens which may be presented.

In accordance with the above direction, provision is now made to preserve and display a collection as fast as it shall be formed. By permission it will be located for the present in the new apartment devoted to the cabinet of Col. Gibbs, and of Yale College.

It is understood to be the wish of the Society, that its members and others would forward specimens illustrative of American Geology and Mineralogy. The names of the donors will be duly recorded, and their donations will be properly acknowledged.

AUTHOR CITATION INDEX

Adams, F. D., 5, 277
Adrain, R., 191
Akerly, S., 277
Albritton, C. C., 5
Alden, T., 12
Aldridge, A. O., 6
Annan, R., 71
Aristotle, 81
Atwater, C., 71, 277

Baker, H., 71
Barton, B. S., 12, 71, 94, 98, 111
Beck, T. R., 278
Beckwith, J., 227
Bell, W. J., 6
Biot, J. B., 191
Bowditch, N., 215
Brickell, J., 227
Brongniart, A., 71
Bruce, A., 150
Buel, D., Jr., 277
Buffon, G. L. L., Comte de, 76, 78, 81, 83, 85, 86
Bullock, W., 80
Burke, J. G., 6, 150

Camden, W., 117
Cavendish, H., 191
Channing, W., 150
Clay, J., 191
Cleaveland, P., 150, 227
Collinson, P., 71
Columbian Magazine or Monthly Miscellany, 277
Cooper, T., 111
Cutbush, J., 150
Cuvier, G., 70, 71, 72

Dana, J. F., 122, 277
Dana, S. L., 277
Daubenton, L. M., 76, 78
Desaguliers, J. T., 45, 58

Dewey, C., 277, 278
De Witt, B., 150, 151
Dijon, 162
Donaldson, 119
Doolittle, T., 12
Drake, D., 111, 278
Dudley, P., 12
Dunbar, W., 72

Eaton, A., 150, 277, 278
Edinburgh Review, 336
Edwards, T., 72
Engelstrom, see Engestrom, G. von
Engestrom, G. von, 152
Evans, L., 278

Fitzpatrick, T. J., 6
Forster, J. R., 91
Fourcroy, A. F. de, 118
Frondel, C., 6
Fulton, J. F., 6, 227

Gadd, P. A., 281
Gibbs, G., 151
Greene, J. C., 6, 71, 150, 335
Griscom, J., 122
Griswold, S., 12
Godon, S., 150

Hall, C. R., 6
Hare, R., Jr., 150, 166
Harpe, de la, 82
Harriot, T., 80
Hawkins, J., 80
Hayden, H. H., 278
Hazen, M. H., 6
Hazen, R. M., 6, 7, 277
Hindle, B., 6, 12
Hitchcock, E., 277
Holly, H., 227
Hosack, D., 111, 112, 122
Houseman, 118

Author Citation Index

Hunter, W., 72, 95

Jefferson, T., 72, 96

Kain, J., 278
Kalm, P., 91
King, E., 264
Kingsley, J., 227
Kirwan, R., 281

Lane, E., 344
La Place, P. S., 215, 216, 217, 218, 219
Latrobe, B. H., 227
Lea, I., 150
Lewis, Z., 227
Linnaeus, 120
Lowthorp, 43, 55

McAllister, E. M., 6
Maclure, W., 256, 278
Magellan, J.-H., 152
Mansfield, J., 191
Martin, M., 115
Mather, I., 12
Meade, W., 122, 126, 134
Mease, J., 278
Meisel, M., 6
Meriwether, D., 72
Merrill, G. P., 6, 227, 277, 335
Mitchill, S. L., 12, 111, 278, 304
Moore, J., 6

Newton, I., 55, 64

Penick, J., Jr., 6
Peale, R., 71, 72
Phillips, W., 151
Pierce, J., 278
Pliny, 81

Pomerantz, M. A., 6
Pott, J. H., 281

Rafinesque, C. S., 278
Rouelle, J., 112
Rudwick, M. J. S., 6

Saumur, 118
Saussure, H. B. de, 281
Say, T., 72
Schneer, C. J., 6, 191
Schoolcraft, H. R., 278
Schöpf, J. D., 278
Seaman, V., 112, 122
Seybert, A., 151
Silliman, B., 151, 227
Simpson, G. G., 6
Spieker, E. M., 7, 277
Steel, J. H., 123, 134
Stevens, J. W., 227
Swinburne, C. H., 119

Tenney, S., 112
Thomson, E. H., 6, 227
Tilloch, T., 183
Turner, G., 72

Virgil, 41
Volney, C. F. C. B., 278

Waterhouse, J. F., 151
Webster, J. W., 278
Wells, J. W., 7
West, S., 72
White, G. W., 7, 227, 277
Williams, S., 12
Winthrop, J., 12
Wistar, C., 72, 86
Woodhouse, J., 227

SUBJECT INDEX

Accum, F., 148–149
Adrain, R., 190
Aikin, A., 172–173
Alabama, 287, 289, 294, 308
Alluvial formation. See Formation, alluvial
American Academy of Arts and Sciences, 190–191
American Geological Society, 5, 335
  activities, 343–346
  cabinet, 346
  Constitution, 341–342
  officers, 149, 343
  organization, 5, 335, 341
American Journal of Science, 150, 224, 225
American Mineralogical Journal, 3, 149, 171, 336
American Philosophical Society
  activities, 166
  Museum, 86, 106, 171
  officers, 68–69
  Transactions, 86, 148, 189, 274, 275, 306, 336
Amherst College, 276
Analyses, chemical, 196
  basalt, 234–235
  minerals, 146, 148–149, 152–164, 167–168
  mineral water, 111, 126–137
Andalusite, 293
Antimony, ores of, 340
Aragonite, 293
Arsenopyrite, 293
Asbestos, 154. See also Serpentine
Assaying. See Analyses, chemical

Baldwin, E., 184
Barbados, earthquake, 52–53, 58, 60
Barite, 90, 178, 292, 293, 296
Barton, B. S., 2, 70, 87–100
Barytes. See Barite
Basalt
  analyses, 235–236
  identification, 170, 185, 208

  localities, 256–257, 297
  North Carolina dike, 223–224, 230–242
  origin, 3, 210, 223, 228–229
Beckwith, J., 224
Bergman, T. O., 235–236
Bermuda, medical topography, 120
Beryl, 344
Bible
  quotations from, 14, 17, 19–24, 30–36
  reference to, 68, 188, 276, 313
Bismuth
  analysis of ore, 158
  Connecticut, 183, 344
Blowpipe
  Hare's 148, 166, 346
  mineral identification with, 152–161, 178–179, 182–183
Blumenbach, J. F., 88
Bones. See Fossils, vertebrate
Boston, earthquakes, 14–16, 39–40, 43–51, 53, 60. See also Massachusetts
Bournon, Count de, 179
Bowditch, N., 3, 190–191, 215–219
Bowdoin, J., 189, 197
Bowdoin College, 70, 101
Boyle, R., 27, 54
Breccia, 119, 212, 228
Brochant de Villiers, A., 147, 184, 186, 337
Brongniart, A., 147, 172–175, 180, 186, 337–338, 344
Brown, S., 343
Bruce, A.
  editor of *American Mineralogical Journal*, 1, 3, 171, 249, 336
  mineral collection, 149, 171
Brucite, 183
Buffon, G. L. L., Comte de
  origin of the earth, 84, 189
  stature of American animals, 69, 84–86
  vertebrate fossils, 76, 78, 81
Bürg, J. T., 218

349

Subject Index

Calcite, 152, 178
Canada
   earthquakes, 50
   fossil elephants, 91
   geology, 287, 298, 339
Carver, J., 273
Causes
   First Cause, 10, 14, 17–25, 28–36, 41–42, 63, 65
   secondary cause of earthquakes, 10–12, 14, 25–28, 51–65
Cavendish, H., 203, 205
Caves
   England, 26
   fossils from, 74
   Indiana Territory, 183
   nitre from, 90
   relationship to earthquakes, 25–27, 52–53, 57–58, 190, 205–206
Cerussite, 178
Chalcopyrite, 178, 291
Chalk. *See* Formation, chalk
Chauncey, C., 51
Chemistry. *See also* Analyses, chemical
   application to mineralogy, 148
   courses, 3, 148, 223
Chlorite, 208
Chromite, 340
Chrysoberyl, 344
Clap, T., 267–269
Classification
   minerals, 3, 146–150, 172–176
   rocks, 222–225, 243–253
Clay
   localities, 102–103, 113, 118, 125, 126, 281, 288, 292, 297
   uses, 280, 308
Clay, J., 190
Cleaveland, P.
   American Geological Society, 335, 343
   Elementary Treatise on Mineralogy and Geology, 3, 5, 70, 147, 149–150, 169–186, 224, 276, 334–340
   fossil shells, 2, 70–71
Coal
   extent in North America, 297, 308, 338–339
   Franklin on, 192
   Maclure on, 258
   Pennsylvania, 211–212, 223
   Rhode Island, 211–212, 296
   Virginia, 292
Cobalt
   analysis of ore, 158
   localities, 294, 340
Collections, mineral, 3, 149, 171, 182, 346

Colleges
   Amherst, 276
   Bowdoin, 70, 101
   Columbia, 113
   Harvard, 11–12, 38, 171
   Princeton, 225
   Queen's, 149
   University of Pennsylvania, 166, 223
   William and Mary, 89
   Yale, 149, 176, 225, 226, 267, 335, 341, 343, 346
Columbia College, 113
Comets
   effects on poles of the earth, 198
   relationship to meteorites, 226, 267–270
   role in formation of earth, 84
Connecticut
   earthquakes, 50
   geology, 254–258, 286–287, 289–291, 294, 308, 311–312, 338, 340, 343–344
   minerals, 176, 183, 184, 343–344
   Weston meteorite, 4, 226
Conrad, T. A., 71
Cooper, T., 1, 12, 190, 201–214, 223
Copper
   analysis of ore, 158–159
   localities of ore, 291, 339
Coxe, J. R., 148
Cronstedt, A., 3, 147–148, 152, 234
Cross-sections, geological, 5, 276, 308, 311
Crystallography, 146, 177
Cutbush, J., 130–131, 134
Cuvier, G.
   Essay on the Theory of the Earth, 2, 70, 184
   extinction, 70, 213
   letter to, 87–100
   Rafinesque on, 307

Dana, J. D., 148, 275
Dana, J. F., 122, 130–134, 275
Dana, S. L., 275
Daubenton, L. M., 75, 76, 78
Davy, H., 269, 312
Day, J., 4, 226, 259–270
De La Metherie, J. C., 184, 307
Delaware, 285–287, 290
De Luc, J.-A., 184
De Manet, Dr., 76
Derham, W., 38
Descartes, R., 188–189
De Witt, B., 149
Diamond, 171
Dike, mistaken for wall, in North Carolina, 223–224, 230–242
Dolomieu, D. G. S. T., 255
Dolomite, 286, 291–292, 295

Subject Index

Drake, D., 310
Dwight, T., 225

Earth
  central heat, 54–58, 193–195, 202–204, 244–245
  density, 203–205
  figure, 3, 190–191, 215–219, 244
  formation, 3, 84, 188–189, 194, 201, 244–245, 312–313
  interior, 3, 189–190, 192–195, 198–199, 243–244
  magnetic field, 3, 189–190, 194–195, 198
  polar wandering, 3, 189–190, 194–195, 198–199, 307
Earthquakes, 2, 9–65. See also Causes; Caves; Sermons
  locations
    Barbados, 52–53, 58, 60
    Lisbon, Portugal, 52–53, 58
    New England, 10–11, 14–16, 38–39, 43–51, 53, 58, 60–61, 65
    Port Royal, Jamaica, 2, 43
    Saint Martin's West Indies, 49, 52–53, 58–59
    Sicily, 43
  origin, 10–12, 17–36, 189–190, 195, 199–200, 205–206
  wave-like propagation, 11–12, 40, 43–49, 53, 58
Eaton, A.
  American Geological Society, 335
  Index to the Geology of the Northern States, 5, 276, 310–313
  on progress in geology, 334
  teaching, 1, 276
Edinburgh Review, 5, 256, 334, 336–340
Elephant. See also Mammoth; *Mastodon*
  extinct species, 69–70, 87–100
  nomenclature, 88
Elliot, S., 343
Emporium of Arts and Sciences, 190
England
  cave, 26
  chalk, 113–116
  coal, 192
  medical topography, 113–118
Epidote, 180, 293
Erosion, 281–282
Evans, L., 273
Extinction
  Barton on, 70, 87, 92–94, 99
  Cuvier on, 70
  Jefferson on, 69, 79, 83–84

Feldspar, 152, 170, 207–208, 280, 292, 344

First Cause. See Causes
Fish, fossil, 338
Fisher, A. M., 343
Flint, 113, 202, 288
Florida, 116, 285, 288
Flötz. See Formation, secondary
Fluorite, 124, 183, 343–344
Fluor Spar. See Fluorite
Formation
  Alluvial Formation, 202, 213–214, 223, 274–275, 284, 287–288, 290, 301–305, 308, 323–330
  Chalk Formation, 110, 113–117, 258, 288
  definition of, 299
  Primitive Formation, 123, 202, 207–209, 211, 212, 223, 256–257, 273, 281, 284–294, 296, 301–302, 304–305, 307–308, 310–312
  Secondary Formation, 202, 208, 211–213, 223, 256–258, 273–274, 284, 287, 290, 294, 296–298, 307
    fossils, 105–108
    mineral springs, 123–124
  Superincubent Formation, 311–312
  Transition Formation, 202, 209–211, 223, 257, 284–285, 287–288, 290, 292–296, 307
Fossils. See also *fossil names*
  implications of, 199
  invertebrate, 2, 70–71, 90, 101–108, 119–120, 124, 192, 209–210, 212–213, 287, 303–304
  plant, 211–212, 292–293, 297, 304
  queries on, 323, 327, 330–331
  vertebrate, 2, 68–70, 73–100, 213, 338
France, 118
Franklin, B., 1, 3, 12, 121, 189, 192–200

Galena, 178, 183, 291–292, 297
Garnet, 153, 171, 178, 280, 292, 293, 343
Gas, natural, 298
Gay-Lussac, J. L., 136
Geological maps..See Maps, geological
Geological Society of London, 314, 330
Geology. See *geographical entries for data*
  early use of the word, 273
  Maclure on object of, 299–300
  Silliman's definition of, 255
Georgia, 285, 287–288, 294–295
Gibbs, G.
  American Geological Society, 335, 341, 343, 346
  mineral collection, 3, 149, 171, 182, 346
Gilmor, R., Jr., 343
Gneiss, 102, 207, 286–287, 292, 293, 302, 343
Gold, 156, 294

351

Subject Index

Granite, 102, 170, 184, 202, 207, 273–274, 280–281, 296, 301–302, 304, 307–308, 343
Graphite, 124, 211
Gravel, 102, 118, 125, 126
Green, J., 71
Greenstone, 290–291
Greywacke, 257, 291–292, 295, 296
Griscom, J., 122–123
Guettard, J. E., 273
Gypsum, 116, 119, 178, 211, 256, 297

Halite, 211, 297
Hall, J. (of New York), 71
Hall, Rev. J., 223, 230–242
Hall, Sir J., 203
Hare, R., 148, 166, 343, 346
Harvard College, 11–12, 38, 171
Haüy, R. J., 147, 177, 180, 186, 337
Hayden, H., 5, 276–277, 314–331
Hazard, S., 106
Hitchcock, E., 1, 5, 184, 275–276, 335, 343
Hornblende, 154, 170, 180, 185, 208, 293–294. See also Schorl
Hosack, D., 122, 149
Hot springs. See Springs, hot
Humboldt, A. von, 255
Hutchinson, T., 273
Hutton, J., 4, 184, 202–203, 224, 307

Iceland, 41
Illinois, 183
Indiana Territory, 183, 224
Indians
  on mineral springs, 124
  myths regarding large mammals, 80–81, 91
  traditions regarding Long Island, 282
Ireland, Giant's Causeway, 223
Iron
  assaying, 158
  localities of ore, 280, 286, 288, 291, 293, 296–297, 339
Italy, Mt. Vesuvius, 41, 56, 64, 264–265

Jamaica, 2, 43
Jameson, R., 147, 172, 181, 184, 186, 201, 223, 337
Jefferson, T., 2, 69, 74–86, 90, 95–96, 226
Johnson, Sir W., 124–125

Kalm, P., 91, 273
Karsten, M., 147
Kentucky, 96, 106, 183
Kidd, J., 173
King, E., 264

Kingsley, J., 4, 226
Kirwan, R., 133, 147, 172, 186, 206, 281, 337

Laboratory apparatus, 152–164
Lamarck, J. B. P. A. de M., 213
Lane, E., mine, 344
Laplace, P. S., 3, 191, 215–219
Lavoisier, A. L., 147
Lea, I., 71
Lead
  analysis of ore, 156–157
  characteristics of ore, 182–183
  localities of ore, 289, 296–297, 339–340
Leibnitz, G. W. von, 189
Le Sueur, C. A., 2, 71, 105–108
Lewis, Z., 224, 237–242
Limestone
  Alluvial, 287
  medical benefits of, 113–121
  Primitive, 286, 291–292
  Secondary, 105–106, 124–126, 211–213
  Transition, 210–211, 290, 295
  uses of, 2, 110, 309
Lion, fossil. See Megalonyx
Louisiana, 285
Lucas, J. A. H., 147, 337

Maclure, W.
  American Geological Society, 335, 343
  Maclurites, 71, 105–108
  mineral collection, 149
  Observations on the Geology of the United States, 4–5, 106, 171, 186, 224–225, 254–256, 274–276, 283–300, 301–311, 334–340
  on rock nomenclature, 4, 224–226, 243–253
  Silliman on Maclure, 254–256
Maclurites, description of, 71, 105–108
Madison, Bishop, 89, 94
Magnetite, 158, 286, 291, 293, 294, 296. See also Iron
Maine, 70–71, 101, 289, 293
Mammoth. See also Mastodon; Elephant
  American, 69, 70, 80, 87–100
  European, 85
Manganese, ore of, 340
Mansfield, J., 190
Maps, geological, 5, 275
  Cleaveland on, 102
  Guettard, 273
  Maclure, 4–5, 274–276, 283, 299, 301–310, 337
  need for, 144
  Volney, 4, 274

Subject Index

Marble
  from Connecticut, 176, 254–255, 343, 344
  formation of, 203
  identification of, 208
  localities, 119, 280, 286, 292–293
Marl, 116, 119
Maryland, 207, 287, 290, 339–340
Maskelyne, N., 203, 205, 218
Massachusetts
  earthquakes, 2, 10–11, 14–16, 38–39, 43–51, 53, 58, 60–61, 65
  geological map, 275
  geological notes, 287–289, 311–312, 338, 340
  minerals, 183, 184, 345
Mastodon, American, 68–69, 70, 88–100. See also Elephant; Mammoth
Mather, C., 68
Mayhew, J., 51
Meade, J., 122–137, 142–144
Mease, J., 274
Medical Repository, 148
Medical topography, 110, 113–121
Medicine, use of mineral springs in, 2–3 111, 122–144, 280
Megalonyx, 69–70, 74–86, 90
Megatherium, from Paraguay, 86, 93
Mercury, lack of American, 340
Meteorites
  Cleaveland on, 182
  Day on origin of, 4, 226, 259–270
  relationship to earth's core, 190, 205
  Weston, Connecticut, fall, 4, 226, 262–263, 266
Mica, 154, 184, 207–208, 280, 292, 293, 295, 343, 344
Mineral collections. See Collections, mineral
Mineralogy. See also Analysis, chemical; Collections, mineral; Minerals
  nomenclature, 179–181
  textbooks, 3, 147–150, 169–186, 224, 276, 334–340
Minerals. See also species names; Analyses, chemical; Collections, mineral; Mineralogy
  classification, 3, 147–150, 172–176
  identification, 3, 152–164, 182–183
  localities, 183–184, 286, 288–298, 343–346
  physical properties, 177–183
Mineral springs. See Springs, mineral
Mines. See also Ores
  Connecticut, 344
  Franklin, New Jersey, 149, 165–168
  temperature of, 54
Mississippi River, 287, 296–297, 339

Mississippi Territory, 288
Mitchill, S. L.
  American Geological Society, 335
  editor of *Medical Repository*, 1, 70
  on medical topography, 2, 110, 113–121
  on New York geology, 4, 273, 275, 279–282, 310
Molybdenite, 184, 293, 294
Mongez, A., 235–236
Moon
  effects of, on figure of the earth, 216–217
  lunar volcanoes as origin of meteorites, 226, 265–267
Moses, account of the deluge, 276, 313
Mt. Aetna, 41, 56, 64
Mountains, queries regarding, 314–317
Mt. Hecla, 41
Mt. Vesuvius, 41, 56, 64, 264–265
Murray, J., 184

Neptunian System, 3, 4, 223, 228–229, 245–253. See also Wernerian System
New England. See also individual states
  earthquakes, 2, 10–11, 14–16, 38–39, 43–51, 53, 58, 60–61, 65
  geological notes, 5, 276, 308, 310–313
New Hampshire, 39, 51, 286, 339
New Jersey
  geological notes, 286–294, 338–340
  mammoth bones, 97
  minerals, 183
  zinc ores, 165–168, 340
Newton, I.
  on density of earth, 203, 204–205
  on origin of geothermal heat, 54–55, 64
New York
  earthquakes, 50
  fossils, vertebrate, 2
  geological notes, 4, 5, 176, 273–274, 279–282, 287, 290, 293–294, 301–305, 308, 310–313, 338–339
  map, geological, 275
  medical topography, 116
  springs, mineral, at Ballston and Saratoga, 2–3, 111, 123–144
Nicholson, W., 233
Nickel
  analysis of ore, 158
  in meteorites and earth's core, 190, 205
North Carolina
  basalt dike mistaken for wall, 3–4, 223–224, 230–242
  geological notes, 285, 287–289, 293–296, 340
Nova Scotia, 287

Subject Index

Ochre, 280
Ohio, 308
Ohio River, 183
Ores. See also Mines
   antimony, 340
   bismuth, 158, 183, 344
   chromite, 340
   cobalt, 158, 294, 340
   copper, 158–159, 291, 339
   gold, 156, 294
   iron, 158, 280, 286, 288, 291, 293, 296–297, 339
   lead, 156–157, 182–183, 289, 296–297, 339–340
   manganese, 340
   mercury, 340
   nickel, 158
   silver, 156, 340
   tin, 156, 340
   tungsten, 344
   zinc, 149, 156, 158, 165–168, 340
Ovid, 169

Paleontology. See Fossils
Pallas, P. S., 90, 98
Patrin, E. L. M., 307, 312
Peale, C. W., 68–69, 73, 97
Pennsylvania
   earthquakes, 50
   geological notes, 207–212, 223, 228–229, 286–287, 289–294, 308, 338–340
   medical topography, 116, 120
   soils, 176
   vertebrate fossils, 2, 73
Periodicals
   American Journal of Science, 150, 224, 225
   American Mineralogical Journal, 3, 149, 171, 336
   American Philosophical Society *Transactions*, 86, 148, 189, 274, 275, 306, 336
   Edinburgh Review, 5, 256, 334, 336–340
   Emporium of Arts and Science, 190
   Medical Repository, 148
Perkins, B., 149, 171
Peru, earthquake, 200
Phillips, W., 173, 184
Pierce, J., 183, 343
Playfair, J., 89, 184, 202–203
Plumbago. See Graphite
Plutonism. See Hutton, J.; Rocks, origin of
Pope, A., 114
Porphyry, 170, 208
Porter, T. D., 343, 345
Portugal, earthquake, 52, 58, 60

Pott, J. H., 152
Prehnite, 184, 291
Pressure, atmospheric, 27
Primary Formation. See Formation, Primitive
Primitive Formation. See Formation, Primitive
Prince, T., 10–11, 13–36
Princeton College, 225
Progress, scientific, 1, 334
Pyrite, 124, 171, 178, 293, 294, 297
Pyrolusite, 184
Pyroxene, 180. See also Schorl

Quartz, 123, 152, 170, 171, 202, 207–208, 280, 290, 294, 344
Queen's College, 149

Rafinesque, C., 1, 5, 71, 275–276, 306–313, 335
Rees, A., 215, 219
Rhinoceros, Siberian fossil, 98
Rhode Island, 211–212, 287–288, 296, 338
Rittenhouse, D., 2, 68, 73
Rivers, queries regarding, 323–331
Rocks. See also names of rocks, i. e. Granite
   classification of, 3, 184–186, 201, 207–214, 222–225, 243–253, 274–276, 283, 311–312
   origin of, 3, 222
Rock salt. See Halite
Rogers, Major R., 273
Rush, B., 148
Rutile, 178, 184, 293

St. Fond, F. de, 235–236
St. Martin's, earthquake, 49, 58–59
Salt licks, fossils from, 92, 96–97, 99–100, 297
Saltpeter, 118
Sand, 102–103, 113, 126
Sandstone, 256–257, 288, 290–291. See also Greywacke
Saussure, H. B. de, 205, 255, 281, 321
Schaeffer, F. C., 343
Schist, 123, 125, 126, 208
Schöpf, J. D., 4, 273
Schorl, 207, 280, 301, 345. See also Hornblende; Pyroxene; Tourmaline
Scotland, medical topography of, 118–119
Scripture. See Bible
Seaman, V., 111, 122–123, 131–132, 134
Secondary cause. See Causes, secondary
Secondary Formation. See Formation, Secondary
Sermons, earthquake, 10–11, 14–36

Serpentine, 183, 208, 293. See also Asbestos
Seybert, A., 148, 149, 171
Shale, 125, 202
Shaw, G., 281
Sicily
    earthquakes, 43
    medical topography, 119–120
    Mt. Aetna, 41, 56, 64
Sienite. See Syenite
Silliman, B.
    American Geological Society, 335, 343–344
    on American mineralogy, 147, 169–173
    on meteorites, 4, 226
    on other scientists, 3, 4, 150, 169–186, 225–226, 254–258
    at Yale, 1, 149
Silver, 156, 340
Slare, Dr., 54, 55, 57
Slate, 207–208, 256–257, 292–293, 295, 296, 343
Sloth. See Megalonyx
Smith, T. P., 3, 171, 223, 228–229
Smith, W., 344
Soapstone. See Talc
Societies
    American Academy of Arts and Sciences, 190–191
    American Geological Society, 5, 149, 335, 341–346
    American Philosophical Society, 68–69, 86, 106, 148, 166, 171, 189, 274–275, 306, 336
    Geological Society of London, 314, 330
Soils
    calcareous, 110, 113–121
    formation of, 309
    nature of, 102–103, 337
South Carolina, 285, 288, 293
Springs
    chalybeate, 280
    gas, natural, 183, 298
    hot, 54
    mineral, 2–3, 90, 111, 122–144
    queries regarding, 326
    salt, 183, 297
Stalagmites, 119
Staurolite, 293, 343
Steatite. See Talc
Steel, J., 123–125, 128–142
Stibnite, 178
Svanberg, L., 219
Sweden, medical topography of, 120
Syenite, 208

Talc, 119, 178, 208, 344

Tennessee, 96, 287, 297
Thompson, T., 184
Tin, 156, 340
Titanium. See Rutile
Topaz, 181
Tourmaline, 293, 345. See also Schorl
Transition Formation. See Formation, Transition
Trap. See Basalt
Tremolite, 293, 343
Tungsten, 344

University of Pennsylvania, 166, 223

Valleys, queries regarding, 318–323
Varen, B., 26
Vermont, 184, 289
Virginia
    fossils
        invertebrate, 90
        vertebrate, 2, 69–70, 74–86, 88–100
    geological notes, 287, 290, 292–294, 296, 298, 308, 338, 345–346
    mineral springs, 2, 90, 111
Volcanic Formation. See Volcanoes, rocks from
Volcanoes, 2, 54
    meteorite origin, 263–267
    none in eastern North America, 280, 286
    relationship to earthquakes, 40–41, 54–58, 64, 195, 205–206
    rocks from, 4, 202, 204, 214, 223–224, 234, 246, 251–253, 254, 307
    temperature of earth, 54–58, 193–195, 202–204, 244–247
Volney, C. F. C., 4, 274, 275, 308–310

Wall, natural, in North Carolina, 3–4, 223–224, 230–242
Waterhouse, B., 171
Webster, J. W., 149, 343
Werner, A. G. See also Wernerian System
    Maclure on, 255, 274, 299, 337
    on mineral classification, 147, 173–175, 177
    on rock classification, 3, 184–186, 201, 207–214, 223, 311–312
    sketch, biographical, 222–223
Wernerian System. See also Neptunian System; Werner, A. G.
    Maclure on, 243–253, 256–258, 283–300, 337
    nomenclature, 4, 184–185, 256–258, 274, 299, 308
    outline by Cooper, 207–214, 223
    Silliman on, 254–256

Subject Index

West Indies
    Barbados, 52–53, 58, 60
    St. Martin's, 49, 58–59
William and Mary College, 89
Williams, S., 11–12
Winthrop, J., 2, 11, 37–65, 198
Wistar, C., 86n
Woodhouse, J., 4, 148, 223–224, 230–242

Yale College
    American Geological Society, 335, 341
    courses in earth science, 176
    faculty, 225, 226, 267
    mineral collection, 149, 343, 346

Zeolites, 155, 156, 180, 183, 291
Zinc
    analysis of ores, 156, 158, 167–168
    localities of ores, 149, 165–168, 340
Zincite, 165–168, 340

About the Editor

ROBERT M. HAZEN, experimental mineralogist at the Carnegie Institution of Washington's Geophysical Laboratory, received the B.S. and S.M. in geology at the Massachusetts Institute of Technology in 1971, and the Ph.D. in mineralogy at Harvard University in 1975. Prior to joining the Carnegie Institution he was a NATO Fellow at Cambridge University in England. His research interests include high-temperature and high-pressure crystallography and the crystal chemistry of rock-forming minerals, as well as the history of North American geology. In addition, as a part-time professional trumpeter, Dr. Hazen has performed with numerous ensembles including the Boston and National Symphony orchestras.